建筑立场系列丛书 No.47

C3

传统与现代
Traditional Modernity

汉英对照版
(韩语版第363期)

韩国C3出版公社 | 编

王晓华 曹麟 时真妹 曹硕 张琳娜 刘九菊 周一 蒋丽 | 译

大连理工大学出版社

建筑立场系列丛书 No.47

4 城市交通的转变

- 004 城市交通的转变 _ Andrew Tang
- 008 双子车站 _ Spora Architects
- 022 鹿特丹中心火车站 _ Team CS
- 034 洛格罗尼奥高速列车站 _ Ábalos + Sentkiewicz Arquitectos
- 044 加里波第广场 _ Dominique Perrault Architecture
- 052 白湾游轮码头 _ Johnson Pilton Walker

66 公共空间与城市规模项目

- 066 公共空间与城市规模项目 _ Heidi Saarinen
- 072 巴拉卡度Pormetxeta广场 _ MTM Arquitectos + Xpiral Arquitectura
- 086 辛特拉论坛 _ ARX
- 100 曼福综合馆 _ TSM Asociados

日式城市住宅
108 传统与现代

- 108 日式城市住宅——传统与现代 _ Michele Stramezzi
- 114 K住宅 _ Hiroyuki Shinozaki Architects
- 122 长廊住宅 _ FORM/Kouichi Kimura Architects
- 132 白砂住宅 _ ARAY Architecture
- 142 Okusawa住宅 _ Hiroyuki Ito Architects + O.F.D.A.
- 150 山外小屋 _ acaa/Kazuhiko Kishimoto
- 160 雪松住宅 _ Suga Atelier
- 168 螺旋之家 _ Hideaki Takayanagi Architects
- 176 阳台之家 _ Ryo Matsui Architects

184 建筑师索引

4 Transforming Urban Transportation

004 *Transforming Urban Transportation _ Andrew Tang*

008 Twin Stations _ Spora Architects

022 Rotterdam Central Station _ Team CS

034 High Speed Train Station in Logroño _ Ábalos + Sentkiewicz Arquitectos

044 Garibaldi Square _ Dominique Perrault Architecture

052 White Bay Cruise Terminal _ Johnson Pilton Walker

66 Public Spaces and Urban Scale Projects

066 *Public Spaces and Urban Scale Projects _ Heidi Saarinen*

072 Pormetxeta Square in Barakaldo _ MTM Arquitectos + Xpiral Arquitectura

086 Forum Sintra _ ARX

100 Mapfre Complex _ TSM Asociados

Japanese Urban Dwell
108 Traditional Modernity

108 *Japanese Urban Dwell – Traditional Modernity _ Michele Stramezzi*

114 House K _ Hiroyuki Shinozaki Architects

122 Promenade House _ FORM/Kouichi Kimura Architects

132 Shirasu House _ ARAY Architecture

142 Okusawa House _ Hiroyuki Ito Architects + O.F.D.A.

150 Beyond the Hill _ acaa/Kazuhiko Kishimoto

160 House of Cedar _ Suga Atelier

168 Life in Spiral _ Hideaki Takayanagi Architects

176 Balcony House _ Ryo Matsui Architects

184 Index

城市交通的转变
Transforming Urban

随着科技的发展和运输速度的提高，车站和运输枢纽的角色变得越发复杂，亟待改善它们的需求客观上创造了机会，让人们重新思考车站的分类以及车站与城市和公共区域的关系。公共休息室、零售区、交流区、城市的标志区和重大活动的举办场地在过去的交通枢纽中都体现得不太明显。现在的交通急需实现与其他交通模式的有效连接，随着迈向信息时代社会的速度的加快，更好地建设基础设施的要求也在增长。而随着预计的人口增长和全新规划的运输网络，城市交通建筑经历了变革，一方面需要升级它的功能性，另一方面需要以21世纪的新精神来重新建立。城市交通新建筑如何展示自己，对改变城市的形象、加强城市的连接功能以及未来的发展方面有很大的影响。

交通运输的基础设施是如何定义我们的城市结构的？公路、自行车道和人行道是最基本的公共区域，它们支撑着我们的运输方式，同时定义着城市的总体格局。在宏观水平上，公路、航道和领空决定着城市区域如何连接彼此的空间。城市规划观点中认为宏观和微观的结合点是交通枢纽、港口和火车站，它们将不同的交通模式结合在一起。机场通常被规划在城市核心之外，而游轮码头可以靠近城市中心，重要的火车站或地铁站通常规划在城市肌理的腹地，即使不是在中心，也要决定着城市的重要中心和发展。

As technology and the speed of transportation improve, the role of the stations and transportation hubs has become more complex, and the need to improve them creates opportunities in rethinking the typology of the station and its relationship with the city and the public realm. Lounges, retail, communication, branding of cities and even event spaces are programs that were less evident in the historic transportation hubs. Better efficiency to connect to other modes of transport is desired, and thus a need for better infrastructure is growing as the speed of our information age society increases. As projected population unfold and the new transport network is proposed, the architecture of urban transportation is going through an evolution to upgrade its functionality while at the same time reestablishing itself in the spirit of the 21st century. How the new architecture of urban transportation manifests itself is having effect in the transforming of the city's image, urban connectivity and future development.

双子车站_Twin Stations/Spora Architects
鹿特丹中心火车站_Rotterdam Central Station/Team CS
洛格罗尼奥高速列车站_High Speed Train Station in Logroño/Ábalos+Sentkiewicz Arquitectos
加里波第广场_Garibaldi Square/Dominique Perrault Architecture
白湾邮轮码头_White Bay Cruise Terminal/Johnson Pilton Walker
城市交通的转变_Transforming Urban Transportation/Andrew Tang

Transportation

近几年，出现了一种需求，即重建旧的车站来满足人口增长的需要，或者建立新的车站来满足新的路线和科技的需要，像穿过欧洲的高速列车网络。这需要建立更有吸引力和有效的交通枢纽，而这个交通枢纽或者建在原来城市的腹地，或者形成一个新的交通枢纽，来创造一个新的机会，这样不仅建造了一座新的建筑，重建城市受欢迎的形象，而且能在再次促进增长的同时改造周边的公共领域。这个交通枢纽在地区水平和当地水平之间重新建立了连接性和城市连贯性。重新思考车站的类别，这一机遇解决了处理大型建筑物的问题，然而具有讽刺性意味的是，在很多地方，大型历史建筑物都被用来区分街区、设置边界，且氛围不受欢迎且不安全。今天，科技允许我们在传统的车站上面和下面建造分层，我们看见埋入地下或抬离地面的火车轨道、增加的多功能商务项目，重新连接的城市。新的车站是多层次的，有清晰的视线和明亮的日光效果，是公共空间的延伸。用作交通枢纽的新建筑是城市改造的催化剂，相对于过去，这展现了今天的数字时代。

位于尼泊尔的由Dominique Perrault建筑事务所设计的加里波第广场是一个基础设施案例，它重新定义了公共空间。这个项目改善了尼泊尔中央车站前原有的主要广场，同时为建造一座全新的地下地铁站留出空间。这处新型代表性公共空间连接着两个车站，围合了有轨电车线路，且建造了停车场，重新规划了交通流线，使广场成为尼泊尔运输系统中最复杂的运输枢纽。在一系列八棵金属制成的人工树下，地下的商业长廊展现了阴影的交织，同时遮盖了连接原有中心车站的几何结构。天篷的穿孔金属板创造不断改变的外观，即三种变化的图案。公共雕像和火车站中间的道路作为主轴将广场分开。围绕地铁站的公共空间具备城市人行道的特征，主路的另一边是带有池塘和其他景观特征的软景花园。

其他城市的交通枢纽通过与水的连接发挥了作用。位于悉尼港、由约翰逊·皮尔顿·沃克设计的白湾邮轮码头是一个便利设施，成为澳大利亚休闲游船项目的延伸。这个设计利用了现存的、古老的桥式起重机

How does infrastructure of transportation define our city structure? Roads, bicycle lanes, and sidewalks are on the very basic level of the public realm that supports our transportation methods while defining the general layout of our city. On a macro level, Railroads, Waterways, and Airspace are spaces that defined how one urban region connects to another. The coming together of the micro and macro on an urban planning perspective is often transport hubs, ports or stations that bring these various modes of transport together. While airports have traditionally been planned outside the urban core, cruise terminals can be placed closer to the center, and major train or subway stations are often planned deeper into the city fabric and define an important focus and development of the city, if not in the center.

There is a need in recent years to renew older stations to accommodate growth in population or build new ones to cater for new routes or technology such as the high-speed trains network throughout Europe. This needs to create a more attractive and efficient transportation hub, either deep in the original urban context or to establish a new one, offering an opportunity to not only create a new building and reestablish the city's welcoming image, but regenerate growth, and transform the public realm around it. It reestablishes the connectivity and urban coherency between the regional level and the local level. The opportunity to rethink the station typology addresses the challenges of the large footprint of a building that ironically in many places have historically divided neighborhoods, set borders, and established an unwelcoming and unsafe atmosphere. Today, as technology allows us to create layers above and below the traditional station, we see train tracks buried or lifted, mix-use commercial programs added, and urban continuity relinked. The new station is multi-leveled, with clear sightlines and daylight, and is an extension of the public space. The new architecture of the transportation hub is a catalyst of urban transformation that is expressive of the digital era of today with respect to the historical one.

The Garibaldi Square in Naples by Dominique Perrault Architecture is an example of infrastructure redefining the public space. The project improves the existing main square in front of Central Station of Naples, while providing a new address to the underground metro station. The new representative public space links the two stations, borders a tramline and includes car parking and the rearrangement of traffic, making the square one of the most complex transport hub in Naples transport system. Under the series of eight metallic "trees", the commercial gallery below ground level enjoys the shadow-play and shading of the geometric structure that relates to the existing central station. The perforated metal on the canopy provides an ever-changing appearance in variations of three patterns. A road as a main axis between the public statue and the train station divides the piazza however. While the public space surrounding the metro station is of an urban pedestrian character, the other side of the main road is made

结构来支撑优雅的波浪形天篷。这个结构原来是20世纪60年代悉尼和欧洲国家之间的集装箱船运流程的一部分，它允许建筑师创造出内部的空间，用灯光来支撑自由、通风、流动的内部空间，勾勒出城市天际线的全景。建筑拥有工业特征，将一个老化的工业场地变成一次有纪念意义的体验，并给城市和广阔的航道网增添了一个迎宾大门。

通过重塑城市交通枢纽来进行的另一个大型城市特征改造在最近于鹿特丹开放的中心火车站得到体现。CS团队，即Benthem Crouwel建筑师事务所、MVSA建筑师事务所以及West 8城市设计与景观建筑事务所合作设计了这一项目。最近十年，这个项目一直都在建设当中，同时每天火车都在正常运行。富有意义的古老车站被重新开发，来满足2025年即将增长到323 000人的客流量。这一增长不但代表重要的交通枢纽变化的需要，而且还是欧洲高速铁路线网的一部分。这个项目的北部是19世纪的住宅区，南部是繁荣的当代城市中心，该建筑将它们与一个商业长廊和一个地面自行车隧道连接起来，同时铁路和站台被提高了一个层次。从实用性方面来看，这个项目解决了多层次交通枢纽的功能复杂性，主要通过清晰且有序连贯的运输序列来完成：火车、地铁、电车轨道、公交车、出租车和汽车车站、停车场，以及5200个地下自行车车位。从建筑学方面来说，这个项目以北部覆盖了全部轨道的玻璃屋顶为特征，这一特征与精致的居住特征相匹配。南部设有一座大厅，带有闪光的雕塑屋顶，契合了周围环境中的现代摩天大楼。为了和城市的历史相连接，来自战后车站的装饰品融入新设计中，包括旧时钟、霓虹字、旗杆以及两个标志着自行车隧道入口的花岗岩雕塑。鹿特丹中心火车站在城市中的影响可以在周围的公共空间中感受到，尤其是车站广场将人行横道引入城市中心，接受来自两边都是Y形人工树的步行大道的人们的致意。人工树与车站内部的钢铁结构相映衬。车站北部的公共区域将现行隧道改道，使其好像仍处在过去的场景中，且靠近入口处，将水、树和绿化带结合在一起，创造出受欢迎的城市形象。一条温暖的、红色花岗岩铺成的车站公共空间像毯子一样从北到南，从内到外地覆盖了车站的公共区域，将原来的港口小镇变成一座友好的世界级港口城市。

Ábalos+Sentkiewicz建筑事务所设计了一个类似的但是更大的城市

up of softer green garden spaces with ponds and other landscape features.
Other cities' transportation hubs are leveraged through its connectivity to water. The White Bay Cruise Terminal in Sydney Harbour by Johnson Pilton Walker is a facility that serves the expanding leisure cruise industry in Australia. The design utilizes the existing historic gantry crane structure to support an elegant wavy canopy roof. The structure was once a part of the container shipping operations between Sydney and Europe in the late 1960's. The use of this structure allows the architects to create an interior space that appears to support a free and hence airy, flexible use of space with daylight framing a panorama of the city skyline. The architecture embraces the industrial character and transforms the fading industrial site into a memorable experience while adding an iconic welcoming gateway to the city and it's extensive waterway network.
Another significant transformation of a city's character through the remaking of its transportation hub can be seen in Rotterdam with its recent opening of the Central Station. Team CS, a collaboration of Benthem Crouwel Architects, MVSA Architects and West 8, designed the project that has been in the making for the last ten years while daily use of the trains was in operation. The previous historically meaningful station had to be redeveloped to cater for the growth of daily travelers of 323,000 by 2025. This growth represents not only a need for a significant transportation hub of the region but also a part of the European High Speed Line network. Located between a 19th Century residential area to the north and a contemporary thriving city-center to the south, the project bridges the two worlds with a commercial passage and a bicycle tunnel on ground level while the rails and platforms are lifted one level above. The project pragmatically addresses the functional complexity of a multilevel transport hub with a clear, coherently organizing array of transportation: trains, an underground metro, trams, buses, taxis and car drop-offs, car park and an underground bicycle parking of 5,200 stalls. Architecturally, the project features on the north a glass roof that covers all the train tracks, matching the delicate residential character, and on the south, it showcases a grand hall with a glimmering sculptural roof that corresponds to the contemporary skyscrapers in its surrounding. To connect with the city's history, ornamentation from the previous post-war station is re-integrated into the new design, such as the old clock, neon letters, flag posts, and two granite sculptures marking the entrances of the bicycle tunnel. The urban impact of Rotterdam Central Station can be felt in the public space surrounding the building, especially with the station square leading the pedestrian into the city center, greeted by people from the promenade on both side rows of Y-shape trees, which correspond to Y-form steel columns inside the station. The public realm to the north of the station will reroute the existing canal closer to the entrance as it were in its history, bringing water, trees and green to the welcoming image of the city. A warm reddish-granite "carpet" like station public space from north to south, inside and out, transforms the once harbor town into a friendly world-class port city.
In a similar kind of program but of a larger urban transformation,

改造项目，即位于洛格罗尼奥的高速列车站，从战略方面来看，它被看做是发展和实现城市总体规划的一个催化剂和起点。这个规划试图重新连接城市的南部和北部，主要通过装饰地下火车轨道、使屋顶成为大型城市娱乐公园的一部分，合并景观和建筑来实现。新建的塔楼规划在公园的松散结构中，同时一些建筑楼群沿着边界来设置，以定义城市的上空空间。建筑的三角外形在车站以及上层的公共空间中始终得到体现。三角形雕塑嵌板的特征将天花板与墙体结构融合在一起，展现了位于上空的洞穴式"景观"，使用户在靠近更深层的地下时，能够增强这种体验。人们应该铭记这座建筑追求的目标，因为它推动了景观式建筑的连续性，将生态性统一到城市规划中。

　　Spora建筑师事务所设计的双子车站是一个改变了布达佩斯的重要基础设施。Szent Gellért tér车站和Fövám tér车站是整个M4地铁项目的一部分。新的车站代表了布达佩斯地下地铁系统的进化，而该地铁系统始于19世纪90年代，终于20世纪90年代。建筑师提出了以21世纪的精神来重新思考这一项目的问题，承认了这一规划空间、结构、建筑和科技的原始目的。其结果是在设计中显示出预先决定的结构的层次性，并在设计中将其吸收。网络状的建筑构件成为建筑的特征。水晶造型的天窗使日光照射到地下，实现了与地上世界的完美融合。Fövám tér车站是一个复杂的交通连接站，在上下层都实现了交通形式的变换。这个火车站试图将公共空间从上层延伸到下层。这样的话，项目实现了在城市交通枢纽区给城市居民创造出熙熙攘攘的共享、生活和旅游的大众空间的机会。

　　交通中心周围的每一个项目都以独特的方式展示出层次性、复杂性，致力于满足新车站不断变化的功能性和多样性的需求。新的基础设施嵌入到城市环境中，不但很好地提高了城市结构的连接性，而且通过上下叠加公共项目来提供双重使用的机会。交通枢纽是一座非常重要的标志性建筑，将人们聚集在一起，且每个项目进而形成了独特的雕塑特征。而近期项目案例的开发创造了进一步发展城市和改善周边环境的机遇。在许多方面，车站允许城市"净化"，并提高公共空间的质量和旅行团体的效率，同时为新城市的观光者创造出具有代表性的抵达体验。

the High Speed Train Station in Logroño by Ábalos+Sentkiewicz Arquitectos is strategically seen as a catalyst and starting point for the new development and realization of an urban masterplan. The plan strives to reconnect the north and south of the city by decking the subterranean train tracks, making the roof an integral part of a wide city recreation park, merging landscape and architecture. New development towers are planned in a loose configuration in the park, while building blocks are planned along the borders to define the urban void. The triangular geometric language of the architecture carries through consistently within the station as well as public space above. The sculptural character of the triangular panels merges ceiling to wall structure, delivering a fresh grotto-like "landscape" above, enhancing the user's experience as one approaches deeper below ground. The ambition of this project should be noted as it pushes the continuity of landscape-like buildings and the integration of ecology in its urban plan.

An important infrastructure project that is transforming Budapest currently is the Twin Stations by Spora Architects. The Szent Gellért tér Station and the Fövám tér Station are part of the overall M4 Metro Line project. The new stations represent an evolution of the underground metro system of Budapest with the first subway project in 1890s and the last project finished in 1990. The architects therefore posed themselves the question of re-thinking the project in the spirit of the 21st century, acknowledging the original intention of the planned spaces, structure, architecture and technology. The result is to reveal the layers of the predetermined structures and incorporate it in their design. Web-like structural members become an architectural feature. The crystal-like skylights bring daylight deep into underground, promoting an integration with the world above. Like the other projects, the Fövám tér Station is a complex traffic junction to interchange one form of transport to another, above and below grade. The stations strive to extend the public space on grade to the ground below, and promote any activities to encourage use of that space. In that way, the project recognizes the opportunity that urban transportation hubs are opportunities to create a bustling common ground for citizens to share, live, and travel.

In their own unique way, each project around the new transportation hub has shown a level of layers and complexity to attend to the functional yet flexible needs of the ever-changing program of the new station. The new infrastructure embeds itself in the urban context that is not only less hindering to the connectivity of the city fabric but offers chances for the double use of that space by stacking public program below or above it. The transportation hub is a significant, iconic building that calls for the coming together of the people and each of the projects has developed its own sculptural language. The development of the recent examples creates opportunities for further city growth and improvement surrounding it. In many ways, the new station allows the city to "clean up" and enhances its public space quality and efficiency for the traveling community while creating a new representative arrival for the visitor of the new city. Andrew Tang

双子车站
Spora Architects

位于布达佩斯的全新的M4地铁线连接了布达的南部和佩斯城的中心（同时也是布达佩斯的市中心），一期工程计划沿着7.34km的线路建造十个车站。在过去的三十年里，布达佩斯的运输方面并没有经过一次大规模的发展。

在20世纪80年代和90年代，M4车站的设计思想和规划反映了70年代和80年代的理念。然而这些车站将要面向安稳跨入21世纪的人们开放。因此，其最具挑战性的目标便是按照原始规划的那样将结构、建筑、技术和空间合理化，同时根据21世纪的精神来重新规划项目理念。项目的一个目标便是鼓励人们使用公共交通工具。建筑师相信车站的建筑质量能够成为人们达成这一目标的工具之一。布达佩斯是一座崇尚折衷主义、浪漫主义以及传统主义的城市，生活在过去。M4则是一处完全不同的世界，是一个地下的世界。因此突出这里是一处公共空间——一处公共地下空间是非常重要的。Szent Gellért tér车站和Fövám tér车站是一对双子车站，都位于多瑙河的河畔。这两座火车站都是由一个（随挖随填的）盒式体量和隧道构成。这个体量由位于不同层次的钢筋混凝土梁进行支撑，最终的结构类似于一张网，如同一个骨架或者骨骼系统一样。

Fövám tér车站

Fövám tér车站不仅仅是一座地铁站，它还是一个复杂的交通枢纽，是电车、公交车、地铁、船只、汽车和行人交会的场所，形成了一处独特的地上和地下公共开放空间。车站是一个全新的多层城市枢纽，是通往布达佩斯古老市中心的门户。利用自然光曾经是建筑师工作的一个重要方面，在Fövám tér车站的表面，建筑师在车站的上方设计了水晶造型的天窗，能够让自然光抵达室内。

Szent Gellért tér车站

这座火车站的盒式体量由位于不同层次的钢筋混凝土梁来支撑。体量的设计由可见的混凝土网格结构来决定。体量为混凝土墙体，被考顿钢覆盖，且设置了两座电梯以及玻璃幕墙，玻璃幕墙面向室内，以在视觉上和物理空间上将部分建筑与表面连接起来。隧道内的墙体和柱子将由马赛克砖所覆盖，极具艺术性，以映射Gellért旅馆使用的Zsolnai瓷砖。

Twin Stations

The new M4 Metro Line planned in Budapest is to connect South-Buda with the city center of Pest which is the heart of Budapest. Ten stations are to be constructed in the first step along the 7.34km-long line. In the last thirty years there has not been such an ambitious development in regards to transport in Budapest. The concept and previous plans for the M4 were made in the 1980s and 1990s with stations reflecting the way of thinking of the 1970s and 1980s. And yet now the architects will open these stations to passengers living well into the 21st century. Thus the most challenging aim was to rationalize the structures, architecture, technology and space as originally planned while at the same time re-thinking the project according to the 21st century's spirit. One of the goals of the project is to encourage people to use public transport. The architects believe that the architectural quality of the stations can be one of the tools used to get people to do this. Budapest is a city of eclecticism, romanticism, and traditionalism; it is living in the past. The M4 is a different world, an underground world. It is important to emphasize that it is a public space – a public space under the ground. The Szent Gellért tér Sta-

Budapest, M4 metro line

Fövám tér车站
Fövám tér Station

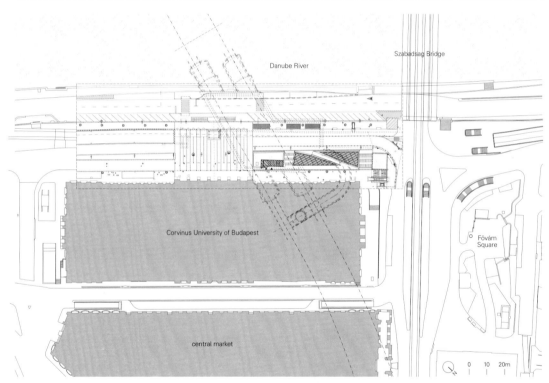

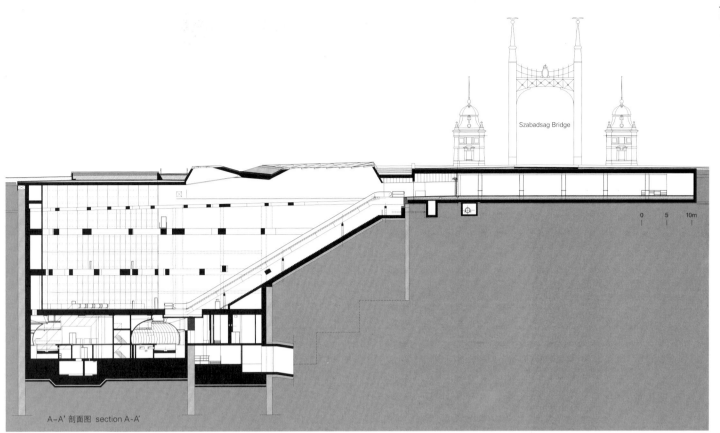

A-A' 剖面图 section A-A'

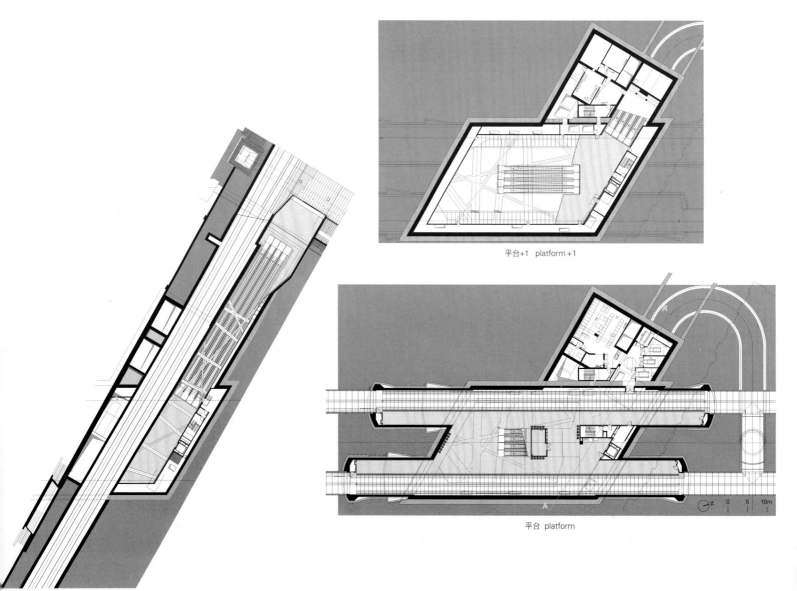

平台+5 platform +5

平台+1 platform +1

平台 platform

tion and the Fövám tér Station are twin stations; both are on the bank of the Danube. They are composed of a cut-and-cover box and tunnels. The box is supported by levels of reinforced concrete beams; the resulting structure is similar to a net, like a bone or skeletal system.

Fövám tér Station

The Fövám tér Station is more than a metro station; it is a complex traffic junction, an interchange spot for tramways, buses, metro, ships, cars and pedestrians, which altogether create a unique open public space above and under the ground. The station is a new multilevel city junction, the gateway to the historic downtown of Budapest. Playing on natural light has been an important aspect of the architects' work; on the surface of Fövám tér the architects designed crystal shaped skylights over the station which let the sunlight reach the interior.

Szent Gellért tér Station

The box of the station is supported by levels of reinforced concrete beams. The design of the box is determined by this visible concrete net-structure. In the main front of the box, which is a concrete wall covered with corten steel, run two elevators with glass walls faced to the inside, in order to connect both visually and physically the parts of the building with the surface. The walls in the tunnels and the columns will be covered with artworks of mosaic tiles reflecting the Zsolnai ceramic tiles of Gellért Hotel.

Spora Architects

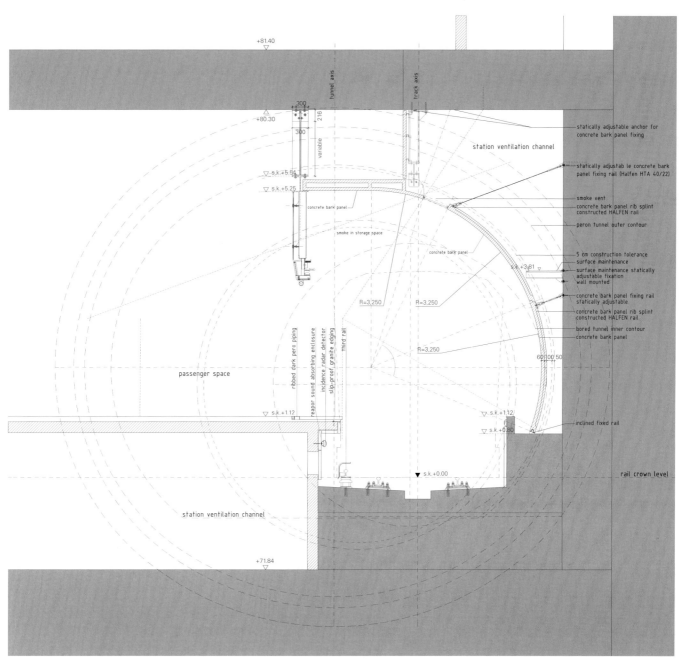

a-a' 剖面图 section a-a'

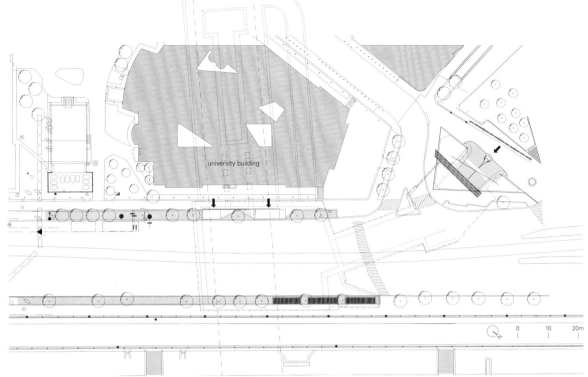

Szent Gellért tér车站
Szent Gellért tér Station

项目名称：M4 Szent Gellért tér / M4 Fövám tér – Underground Station, Budapest
地点：Budapest, Hungary
建筑师：Spora Architects
建筑设计师：Tibor Dékány, Sándor Finta, Ádám Hatvani, Orsolya Vadász
设计团队：Zsuzsa Balogh, Attila Korompay project architects, Bence Várhidi, Noémi Soltész, András Jánosi, Diána Molnár, Károly Stefkó
M4地铁线的总建筑师：Palatium Stúdió
顾问：consortium of Fömterv, Uvaterv, Mott-Macdonald
甲方：Gellért tér_Budapest Transport Ltd. DBR Metro Project Directory / Fövám tér_Budapest Transport Ltd. DBR Metro Project Directory
用地面积：Gellért tér _ 3,000m² / Fövám tér 3,000m²
有效楼层面积：Gellért tér _ 7,100m² / Fövám tér 7,100m²
设计时间：2005—2012
施工时间：2006—2014
摄影师：©Adam Hatvani (courtesy of the architect)

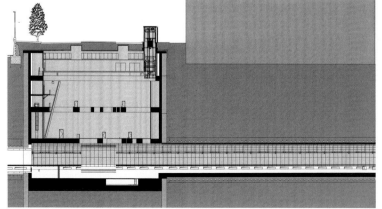

B-B'剖面图 section B-B'

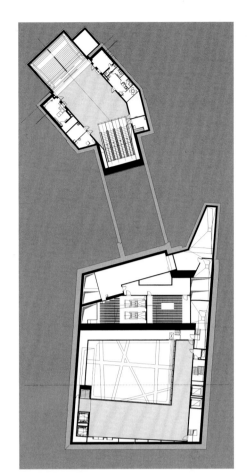

平台+4 platform +4

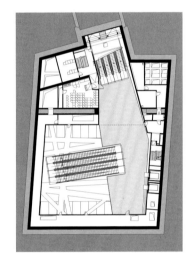

平台+1 platform +1

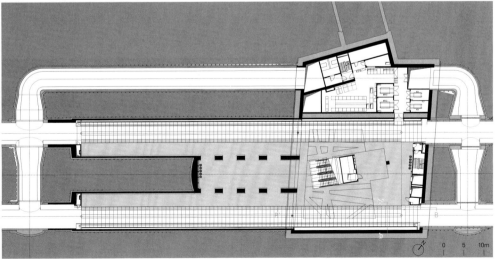

平台 platform

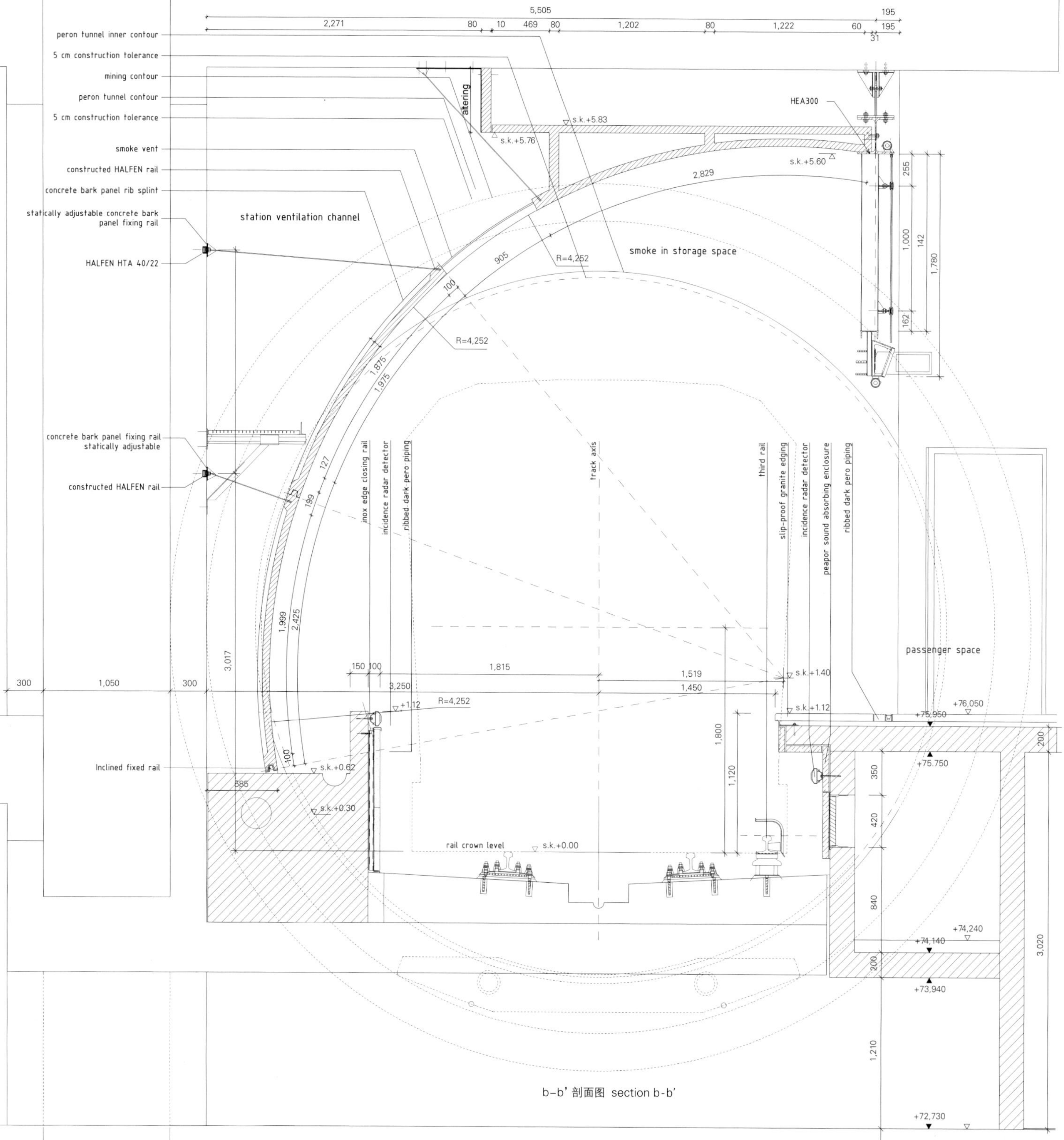
peron tunnel inner contour
5 cm construction tolerance
mining contour
peron tunnel contour
5 cm construction tolerance
smoke vent
constructed HALFEN rail
concrete bark panel rib splint
statically adjustable concrete bark panel fixing rail
HALFEN HTA 40/22
concrete bark panel fixing rail statically adjustable
constructed HALFEN rail
Inclined fixed rail
station ventilation channel
smoke in storage space
HEA300
altering
s.k.+5.83
s.k.+5.76
s.k.+5.60
R=4,252
track axis
inox edge closing rail
incidence radar detector
ribbed dark pero piping
third rail
slip-proof granite edging
incidence radar detector
peapor sound absorbing enclosure
ribbed dark pero piping
passenger space
s.k.+1.40
s.k.+1.12
s.k.+0.62
s.k.+0.30
rail crown level
s.k.+0.00
+76,050
+75,950
+75.750
+74,240
+74,140
+73,940
+72,730
5,505
2,271
1,202
1,222
3,250
1,815
1,519
1,450
3,017
3,020

b–b' 剖面图 section b-b'

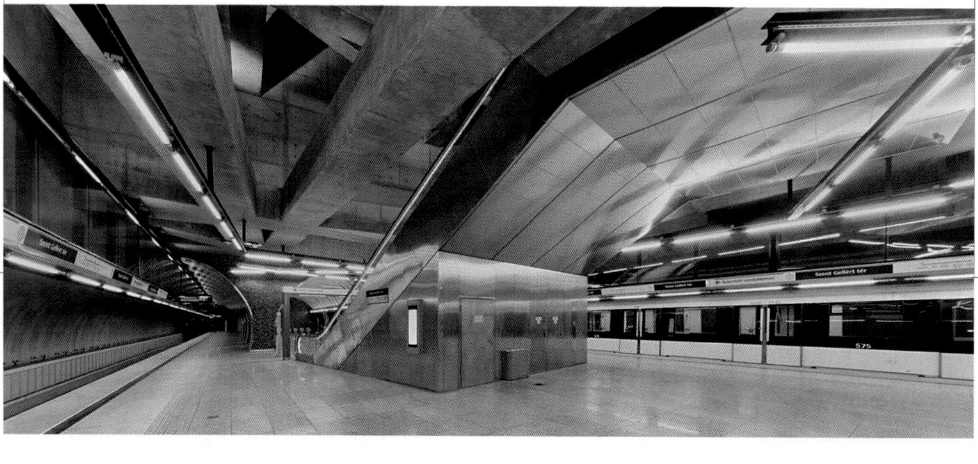

concrete slab

+70,730

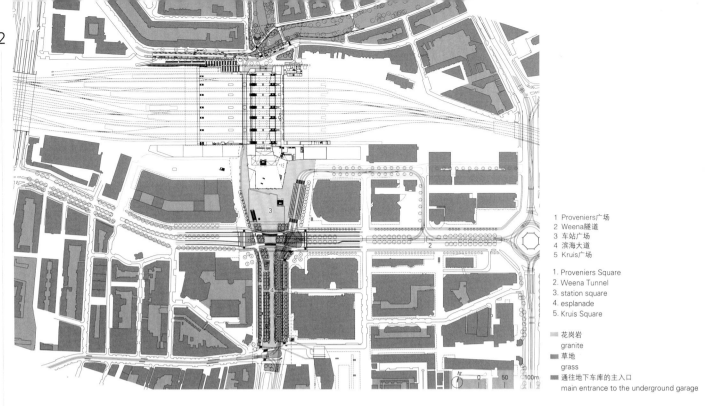

1 Proveniers广场
2 Weena隧道
3 车站广场
4 滨海大道
5 Kruis广场

1. Proveniers Square
2. Weena Tunnel
3. station square
4. esplanade
5. Kruis Square

花岗岩
granite
草地
grass
通往地下车库的主入口
main entrance to the underground garage

鹿特丹中心火车站

Team CS

鹿特丹中心火车站是荷兰最重要的交通枢纽之一。这个公共交通枢纽站拥有与阿姆斯特丹的史基浦机场一样多的客流量,每日高达110000人次。除了与欧洲高速列车网络连接之外,鹿特丹中心火车站也与任仕达铁路的轻轨系统连接。

鹿特丹内的高速列车是荷兰境内从南方去旅行的第一站,它战略性地占据了欧洲的中心位置,距离史基浦机场仅20分钟路程,到巴黎仅仅两个半小时路程。因此,新站不仅比原来的车站更大、更明亮、更有序,而且也更有国际感。其实用性、容量、舒适性及建筑魅力等所有方面都足以与马德里、巴黎、伦敦和布鲁塞尔的中心车站相媲美。

鹿特丹中心火车站面临的一项基本的挑战是车站南北两侧的城市个性差异。北侧入口设计风格内敛,正好配合了相邻的普分尼斯区以及客流量较少的特点。入口逐渐地连接到城市。相比之下,面向城市一侧的宏伟入口显然是通往高层城市中心的大门。

大厅的屋顶完全覆以不锈钢的巨大屋顶,使这座建筑成为城市的地标,并直指城市的中心。

现在的鹿特丹中心火车站的结构与维度适用于城市景观设计。车站前的滨海大道是一处连续的公共空间。为了达到简约的效果,可容纳750辆汽车的小型停车场和5200辆自行车的停车棚建于广场之下。电车站移到了火车站东侧,这样站台的边界拓宽了。

自然光的渗入、太阳光的暖意以及现代化的外观融合成为设计中的重要元素。大厅内部的木质装饰与站台屋顶的木质横梁结构结合,共同营造了一种温暖而热情的氛围,邀请旅客徘徊其间,流连忘返。大面积的透明屋顶结构遮盖了超过250m的全段轨道,将阳光倾洒在站台上。屋顶的玻璃板通过太阳能电池的不同形式来改变透光率,创造了一出在站台上演的、千变万化的、令人着迷的光影剧目。

途径车站的路线是合理的。通过延伸进透明屋顶平台和下到楼梯的空间,从而伸入旅客通道的日光,同直观可见的列车都为旅客提供了指引,拓宽的旅客通道是透明的,其内部嵌入了商业广告,增加了车站的自然一景。

这处公共交通枢纽站为旅客提供的舒适设计在车站的不同地方都可见。它包括商业空间、休息室、餐厅、办公室以及停车场和自行车棚。这里也有旅行信息,即一个信息点、荷兰铁路旅客商店、自动售票机及商业区。在大型咖啡馆和荷兰铁路休息室,整个大厅及临近轨道的壮观景象可尽收眼底。大厅的候车区和通道由客流连接,同时提供阅览和快速购物的区域。

新鹿特丹中心火车站是一处令人愉快的、开放透明的公共交通终端,是一个标志性的交会点。车站交织在城市网络之中,连接着这座城市的不同特性,标记着鹿特丹文化轴线的开端。

在总面积为28 000m²的屋顶上,镶有130 000块太阳能电池的窗户覆盖了10 000m²的区域。考虑到鹿特丹市中心火车站周围的高层建筑,太阳能电池放置在屋顶上,能够最大限度地获得太阳能。玻璃板通过太阳能电池的不同形式来改变透光率。屋顶上电池密度最大的地方,就是通过太阳获得能量最多的地方。

Rotterdam Central Station

Rotterdam Central Station is one of the most important transport hubs in the Netherlands. With 110,000 passengers a day the public transport terminal has as many travelers as Amsterdam Airport Schiphol. In addition to the European network of the High Speed Train(HST), Rotterdam Central Station is also connected to the light rail system, RandstadRail.

Rotterdam HST is the first stop in the Netherlands when travelling from the south and is strategically positioned in the middle of Europe, with Schiphol only twenty minutes and Paris a mere two and a half hours away. Hence the new station is not only larger, brighter and more orderly than the former, but also has an international feel. The station matches in all respects of practicality, capacity, comfort and allure, of the central stations of Madrid, Paris, London and Brussels.

One of the fundamental challenges of Rotterdam Central Station was the difference in the urban character of the north and south side of the station. The entrance on the north side has a modest design, appropriate to the character of the neighborhood Provenierswijk and the smaller number of passengers. The entrance gradually connects to the city. In contrast, the grand entrance on the city side is clearly the gateway to the high-rise urban center. The roof of the hall, fully clad with stainless steel, gives rise to building's iconic character and points to the heart of the city.

Now Rotterdam Central Station has the appropriate structure and dimensions for the urban landscape. The esplanade in front of the station is a continuous public space. To achieve this simplicity a

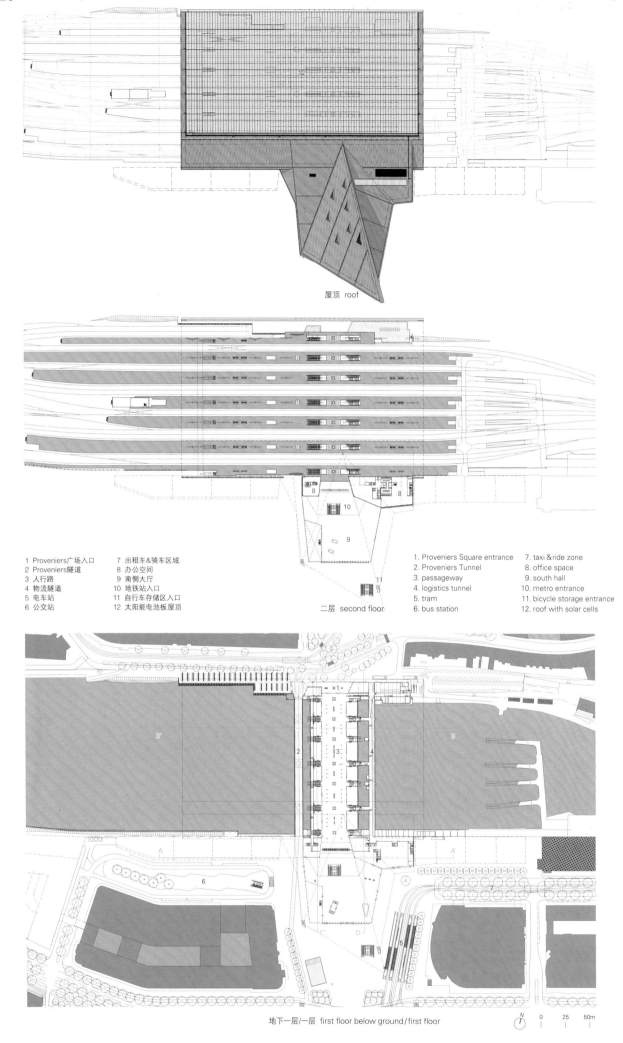

屋顶 roof

1 Proveniers广场入口　7 出租车&骑车区域
2 Proveniers隧道　　 8 办公空间
3 人行路　　　　　　 9 南侧大厅
4 物流隧道　　　　　 10 地铁站入口
5 电车站　　　　　　 11 自行车存储区入口
6 公交站　　　　　　 12 太阳能电池板屋顶

1. Proveniers Square entrance　7. taxi &ride zone
2. Proveniers Tunnel　　　　　 8. office space
3. passageway　　　　　　　　 9. south hall
4. logistics tunnel　　　　　　 10. metro entrance
5. tram　　　　　　　　　　　 11. bicycle storage entrance
6. bus station　　　　　　　　 12. roof with solar cells

二层 second floor

地下一层/一层 first floor below ground/first floor

parking garage for 750 cars and a bicycle shed for 5,200 bicycles are located under the square. The tram station is moved to the east side of the station, so the platforms broaden the square.

Incorporation of natural light, the warmth of the sun's rays and a modern look are important elements in the design. The wood finish on the inside of the hall, combined with the structural wooden beams of the platform roof creates a warm and welcoming ambience, inviting visitors to linger. The largely transparent roof structure which covers all the tracks over a length of 250 meters, floods the platforms with light. The glass plates of the roof vary the level of light transmittance by utilizing different solar cells' patterns, which produce an ever-changing and fascinating play of shadows on the platforms.

The routing through the station is logical; travelers are guided by a direct view of the trains and by the daylight that penetrates to the traveler's passage via the voids that extend through the transparent roof platform and down to the stairs. Because of its transparency the widened traveler's passage, lined with commercial functions, forms a natural part of the station.

The public transport terminal is designed for passenger comfort, which is visible in the different zones of the station. It includes commercial spaces, a lounge, restaurants, offices, parking for cars and bicycles. There is travel information: an information point, the Dutch Railways(NS) travelers shop, ticket vending machines and commercial functions. The grand cafe and the NS-lounge offer spectacular views across the hall and the adjacent tracks. Waiting areas in the hall and the passage are linked to the passenger flows, with areas both for browsing and quick shopping.

The new Rotterdam Central Station is a pleasant, open and transparent public transport terminal which functions as an iconic meeting point. Interwoven into the urban network, the station connects the diverse characters of the city and marks the beginning of Rotterdam's cultural axis.

Windows with 130,000 solar cells cover 10,000m² of the total roof area of 28,000m². The solar cells are placed on the parts of the roof that get the most sun, taking into account the high buildings around Rotterdam Central Station. The glass panels vary in light transmittance by using different patterns in the solar cells. Where the roof has the greatest efficiency in terms of sunlight, the cell density is the highest.

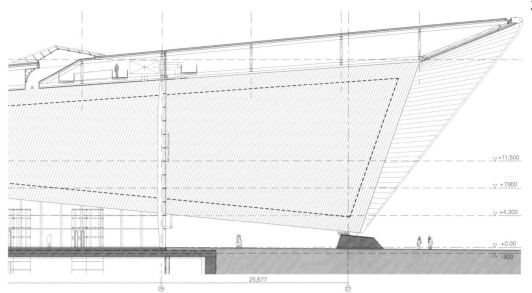

详图1 detail 1

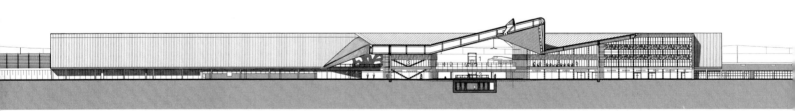

1 南侧大厅 2 地铁入口 3 办公空间
1. south hall 2. metro entrance 3. office space
A-A' 剖面 section A-A'

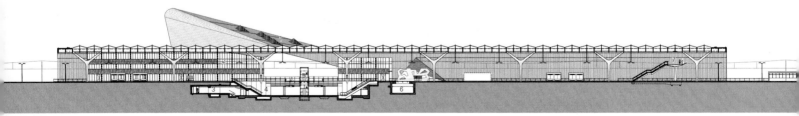

1 漫射自然光的天窗 2 办公空间 3 物流隧道 4 零售空间 5 人行路 6 Proveniers隧道 7 轨道之上的人行桥
1. skylights for diffused natural light 2. office space 3. logistics tunnel 4. retail 5. passageway 6. Proveniers Tunnel 7. footbridge over the tracks
B-B' 剖面图 section B-B'

项目名称：Rotterdam Central Station
地点：Stationsplein 1, 3013 AJ Rotterdam, The Netherlands
建筑师：Team CS _ Benthem Crouwel Architects,
MVSA Meyer en van Schooten Architecten, West 8
首席建筑师：Jan Benthem, Marcel Blom, Adriaan Geuze, Jeroen van Schooten
项目团队：Arman Akdogan, Anja Blechen, Freek Boerwinkel, Amir Farokhian,
Joost Koningen, Joost van Noort, Falk Schneeman, Daphne Schuit, Matthijs Smit(†),
Andrew Tang, Wouter Thijssen, Joost Vos
结构工程师/机械服务工程师/建筑物理工程师：
Arcadis, Gemeentewerken Rotterdam
承包商：Bouwcombinatie TBI Rotterdam Central, Iemants NV
甲方：Gemeente Rotterdam, ProRail
有效楼层面积：46,000m²
有效楼层面积（城市设计方面）：50,000m²
设计时间：2003 / 施工时间：2007 / 竣工时间：2014
摄影师：©Jannes Linders(courtesy of the architect)(except as noted)

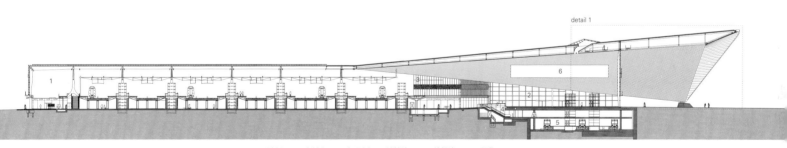

1 北侧大厅 2 南侧大厅 3 办公空间 4 地铁站入口 5 地铁站 6 LED屏幕
1. north hall 2. south hall 3. office space 4. metro entrance 5. metro 6. LED screen
C-C'剖面图 section C-C'

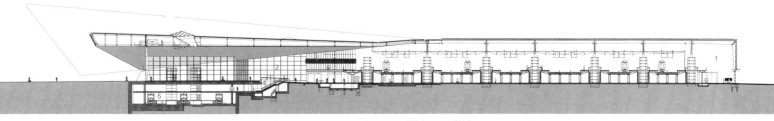

1 北侧大厅 2 南侧大厅 3 办公空间 4 地铁站入口 5 地铁站
1. north hall 2. south hall 3. office space 4. metro entrance 5. metro
D-D'剖面图 section D-D'

洛格罗尼奥高速列车站

Ábalos+Sentkiewicz Arquitectos

这个车站作为一个全新的城市项目的起始点,来重新恢复洛格罗尼奥北部和南部之间的连接。同时,它还附建了一座大型公共花园。车站的屋顶立于公园的中心,并且赋予列车站外形和地理特点。

所有位于街道水平面的车站都会在城市肌理内产生直接的不连续性。但是这个将城市和地面连接且合并的元素却没有产生使城市和社会隔离的孔洞。建筑师重新构思了车站的类型,即掩藏其轨道,这一机遇应该是构思车站的建造方法中的一个转折点。这个联运车站不仅成为改造城市、建造公共空间、开发绿色环带、强化行人和自行车的流动性的机遇,而且还形成了一个突出集体城市体验的全新地形。

从一开始,这个项目便将基础设施、城市规划、景观、建筑、生态和经济问题集在一起,这便是LIF2002项目设计与众不同的地方。换句话说,多方面的管理方式在所有的施工阶段一直在追求质量和创新,并且专注于数量和质量方面。因此,在地形建筑以及生态的都市生活方面,这个车站属于开创性的实验项目。

车站的屋顶结构满足了复杂的隧道轨道的布局需求,形成了一个不规则的四面体布局,同时也满足了维持支撑屋顶木材和混凝土板的承橡梁之间的3m间距的要求,且这个间距是固定的。因此,屋顶也被塑造成山的形状,使其能够沿着平台上面的400m长的水平板表面来创造连续性。

这个结构拥有上下两个表面,且尽量利用两个表面之间的体量,来试图产生可能的、最单一化的布局,如最规律的楼层平面布局。因此,不仅所有桁架的上下部分和公园的表面地形之间具有兼容性,而且底部弦杆和与多面的室内表面相对的柱子之间也具有兼容性。因此,每个网格间的桁架都具有独特的几何外形,与众不同。

在各种技术体系的结构和布局建好之后,建筑师还建造了一个铝质挡板系统。他们专门设计了1.5cm×5cm的铝片,在场地进行组装,带有1.2m×1.8m的标准框架,固定在结构上。这个系统使室内展现在三角形的视野之内,随着人们的移动,其透明性也产生了变化。

该车站的铺装由8cm×8cm的瓷砖铺成,置于玻璃纤维网格中。因为每块砖都进行了不同程度的抛光,因此路人的移动会产生光的变化。项目旨在避免中型和大型设计中的标准地面和天花板的中性特征,同时通过人们的移动使空间被激活。

High Speed Train Station in Logroño

The station serves as a starting point for a new urban project to restore the link between the north and the south of Logroño. At the same time, it gives rise to a large public park. The station's roof stands at the heart of the park and lends its geometrical and topographical qualities to the station's volume.

All street level stations trigger brusque discontinuity in urban fabric. The very element intended to link and merge city and land ends up leaving a gaping hole causing both urban and social segregation. The opportunity to rethink the typology of station

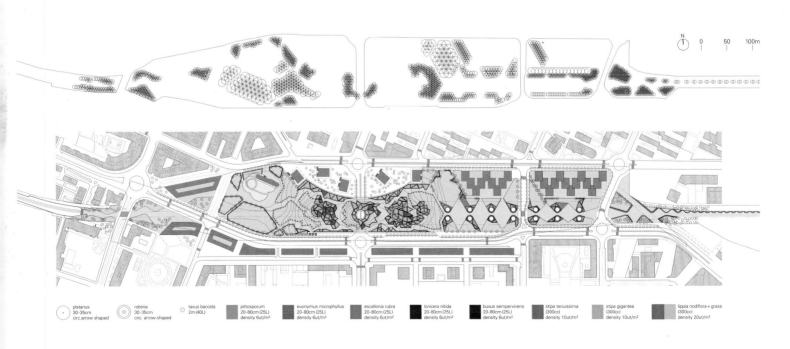

之前 before   之后 after

to which burying tracks gives rise should mark a turnaround in the way stations are conceived. This intermodal station stands not only as an opportunity to transform the city, create public space, develop green rings, and foster both pedestrian and bicycle mobility, but also to create a new topography intensifying a collective city experience.

What makes the LIF 2002 design unique is the fact that from the start, it tackled infrastructure, urban planning, landscape, architecture, ecology and economic issues all together. In other words, its multi-faceted management approach sought quality and innovation throughout all of the phases of the process and attended to both quantitative and qualitative aspects. One therefore can speak in this regard of a pioneering experiment both in landform buildings and in ecological urbanism.

The station's roof structure balances out the need to address the complex track tunnel configuration, and this is done with an irregular tetrahedral configuration, and to maintain the regular three metre spacing between the binding rafters supporting the roof's wood and concrete slab. The roof is thus shaped into a hill providing continuity along the horizontal 400-metre long slab surface above the platforms.

Having established the upper and lower surfaces, the structure designed makes the most of the volume between the two, and

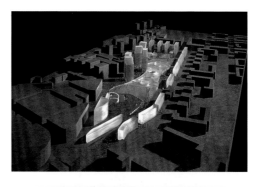

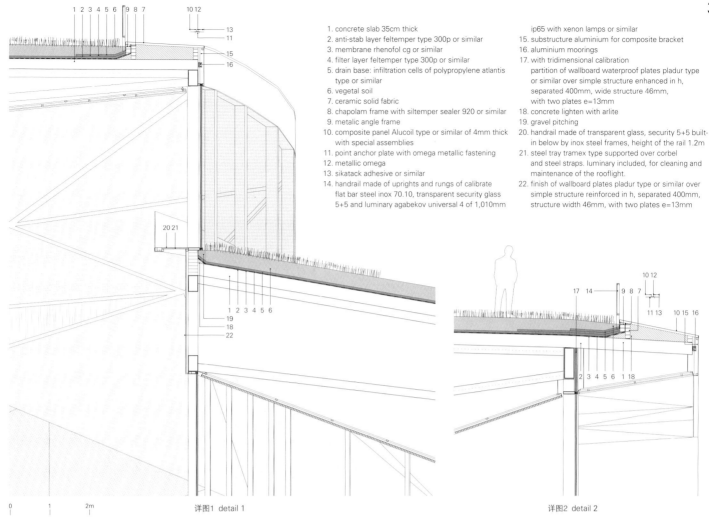

1. concrete slab 35cm thick
2. anti-stab layer feltemper type 300p or similar
3. membrane rhenofol cg or similar
4. filter layer feltemper type 300p or similar
5. drain base: infiltration cells of polypropylene atlantis type or similar
6. vegetal soil
7. ceramic solid fabric
8. chapolam frame with siltemper sealer 920 or similar
9. metalic angle frame
10. composite panel Alucoil type or similar of 4mm thick with special assemblies
11. point anchor plate with omega metallic fastening
12. metallic omega
13. sikatack adhesive or similar
14. handrail made of uprights and rungs of calibrate flat bar steel inox 70.10, transparent security glass 5+5 and luminary agabekov universal 4 of 1,010mm ip65 with xenon lamps or similar
15. substructure aluminium for composite bracket
16. aluminium moorings
17. with tridimensional calibration partition of wallboard waterproof plates pladur type or similar over simple structure enhanced in h, separated 400mm, wide structure 46mm, with two plates e=13mm
18. concrete lighten with arlite
19. gravel pitching
20. handrail made of transparent glass, security 5+5 built-in below by inox steel frames, height of the rail 1.2m
21. steel tray tramex type supported over corbel and steel straps. luminary included, for cleaning and maintenance of the rooflight.
22. finish of wallboard plates pladur type or similar over simple structure reinforced in h, separated 400mm, structure width 46mm, with two plates e=13mm

详图1 detail 1   详图2 detail 2

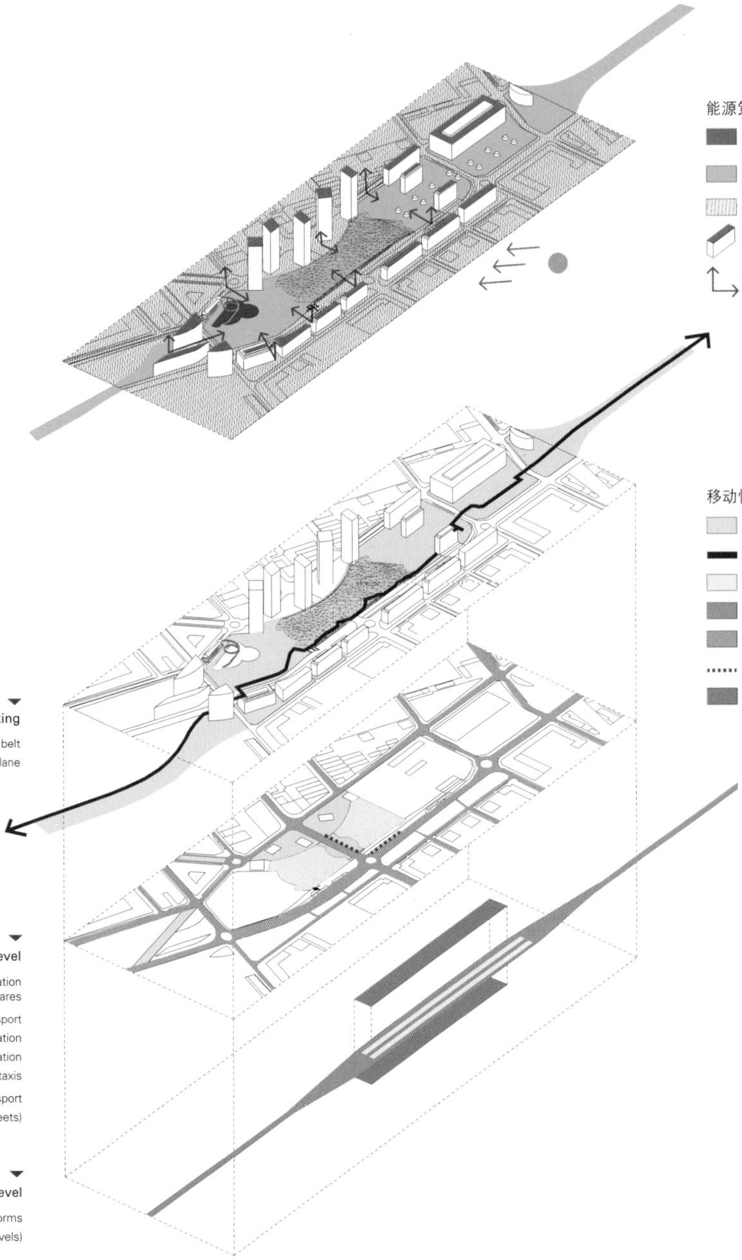

能源策略 energetic strategy
- photovoltaic, thermal & wind collectors (south facing)
- green rooftops. integrated hub station, thermal accumulators & district hot/cold network
- climate improvement of adjacent area
- north-south oriented properties
- capturing rain & underground water (geothermal housing)

移动性 mobility
- pedestrian parks, streets & squares
- bicycle lane
- intermodal train/bus station
- underground track
- urban roads
- taxi
- underground parking

高位停车场 high level parking
pedestrian public park green belt
bicycle lane

一层 first level
pedestrian circulation
streets & squares
public transport
- train station
- bus station
- taxis
private transport
- vehicles (streets)

地下层 underground level
train station platforms
public parking (2 levels)

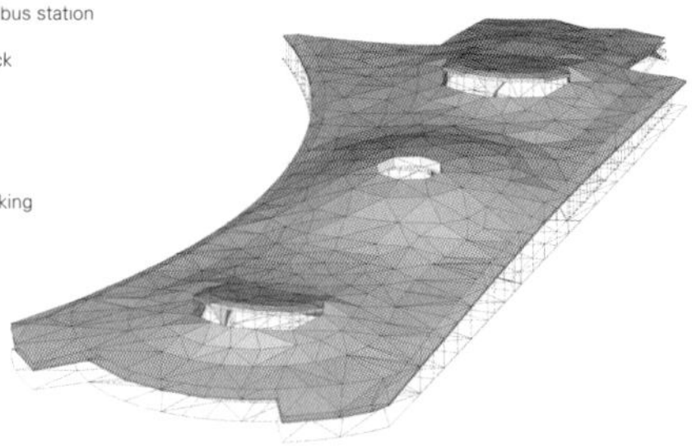

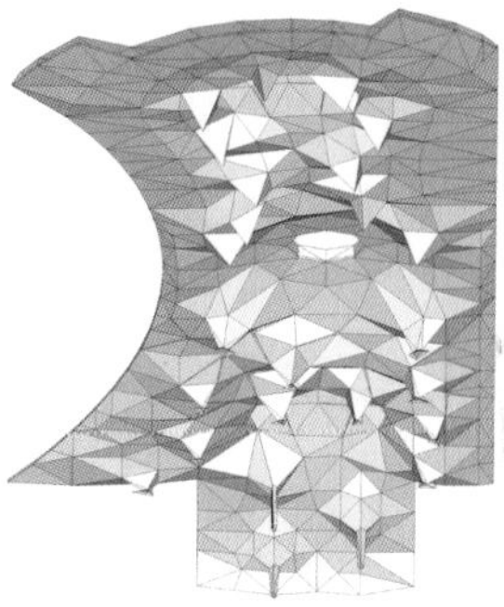

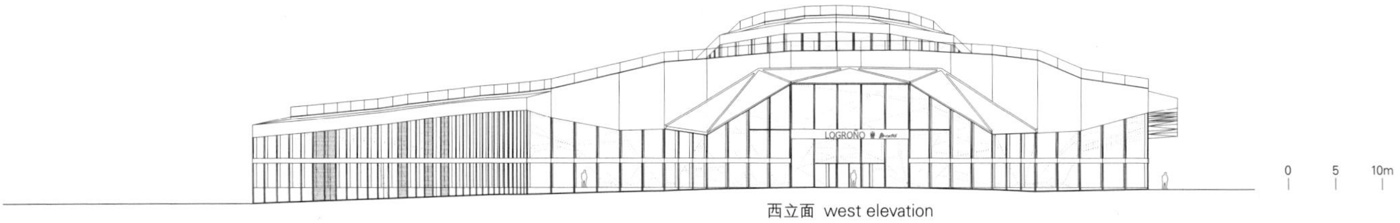

西立面 west elevation

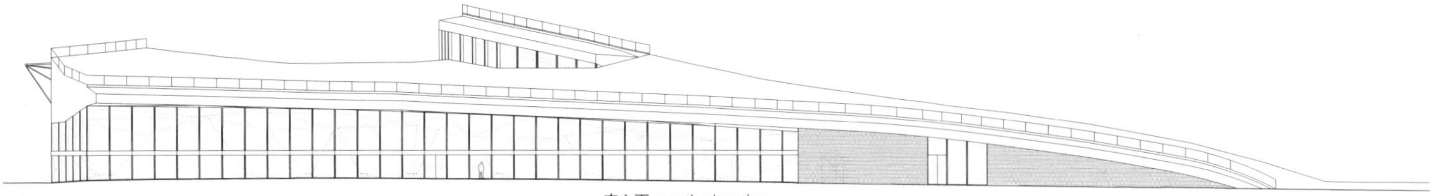

南立面 south elevation

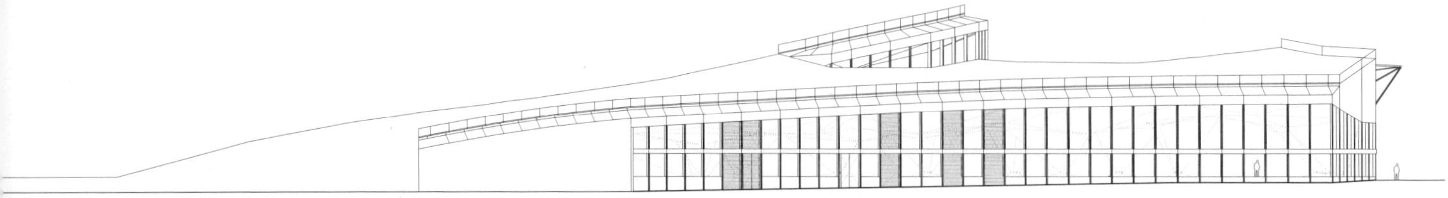

北立面 north elevation

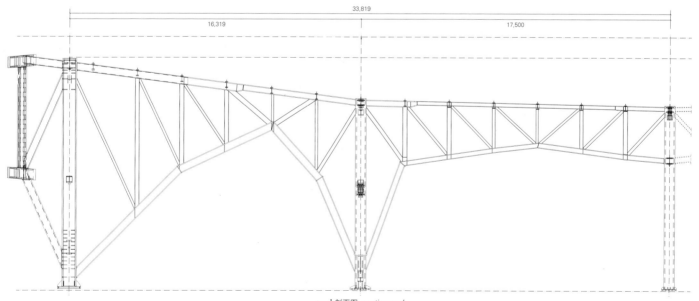

a-a' 剖面图 section a-a'

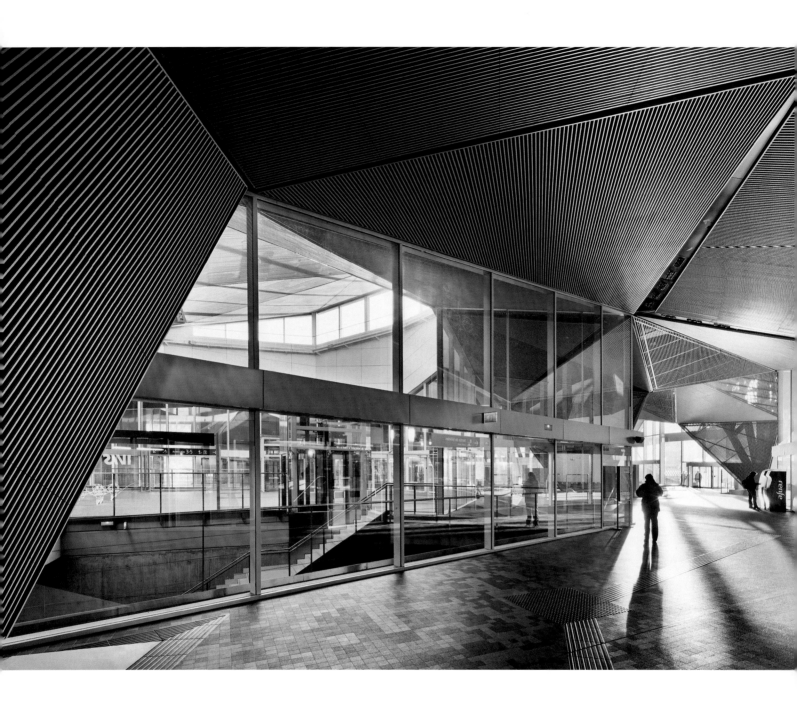

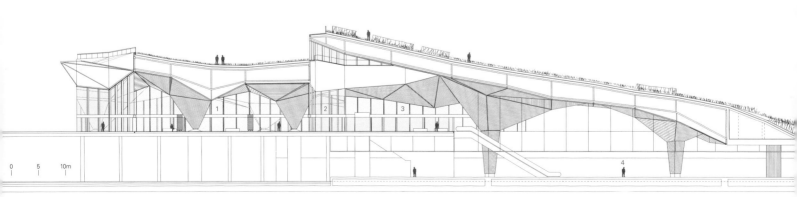

1 入口大厅 2 平台入口 3 等候厅 4 平台
1. entrance hall 2. platform access 3. waiting hall 4. platforms
A-A' 剖面图 section A-A'

attempts to bring about the most homogeneous distribution possible, i.e. the most regular distribution of the floor plan. This thereby generates compatibility not only between the upper and bottom portions of all of the trusses and the park's surface topography, but also between the bottom chords and pillars vis-à-vis the faceted interior surface. As a result, each one of the trusses in the grid affords its own unique geometry, different from the rest. After having established the structure and layout of the various technical systems an aluminium sheet system was built. A patent including 1.5cm × 5cm aluminium sheets is assembled on site with 1.2m × 1.8m standard frames anchored to the structure. The system enables the chamber's content to be revealed in triangulated shots whose degrees of transparency change as one moves. The station's paving was done in 8cm × 8cm gres ceramic blocks on a fibreglass grid. Because each paving block has a different degree of polish, the movement of people as they pass by generates light vibrations. The aim was to avoid the neutral nature of standard floors and ceilings used for medium and large-scale designs, and to enable the space to be activated by the movement of passengers.

项目名称：High Speed Train Station / 地点：Logroño, La Rioja, España
建筑师：Ábalos + Sentkiewicz Arquitectos
项目总监：Francisco Cifuentes, Pelayo Suárez
建筑设计项目总监：Iñaki Ábalos, Alfonso Miguel, Renata Sentkiewicz
项目团队：Jorge Álvarez-Builla, Yeray Brito, Aaron Forest, Victor Garzón, Pablo de la Hoz, Ismael Martín, Laura Torres, Rodrigo Rieiro, José Rodríguez
操作管理：INECO, TYPSA / 建筑管理顾问：Iñaki Ábalos
结构/M&E工程师：UTE Ineco-Sener
景观建筑师：Arquitectura Agronomía, Ábalos + Sentkiewicz Arquitectos
施工单位：SACYR / 甲方：LIF 2002, SA
面积：train station _ 8.000m² / platform _ 19.000m² / parking area _ 18,000m² / bus station area _ 10,800m² / urbanization: 145,000m² / housing area(towers) _ 41,250m² / housing area(other housing) _ 83,750m²
造价：EUR 108,000,000
竞赛时间：2006 / 施工时间：train station _ 2009 / park _ 2009 / bus station _ 2014
竣工时间：train station _ 2011 / park _ 2012 / bus station _ in construction
摄影师：©José Hevia

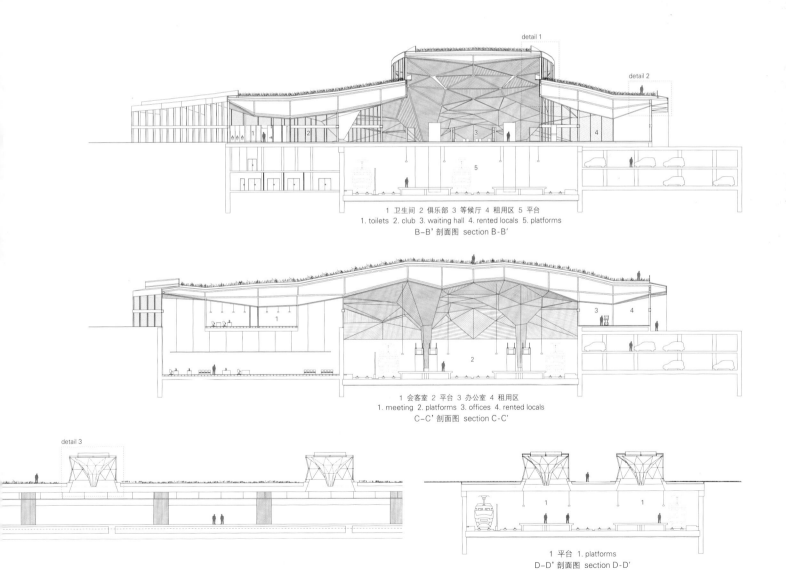

1 卫生间 2 俱乐部 3 等候厅 4 租用区 5 平台
1. toilets 2. club 3. waiting hall 4. rented locals 5. platforms
B-B' 剖面图 section B-B'

1 会客室 2 平台 3 办公室 4 租用区
1. meeting 2. platforms 3. offices 4. rented locals
C-C' 剖面图 section C-C'

1 平台 1. platforms
D-D' 剖面图 section D-D'

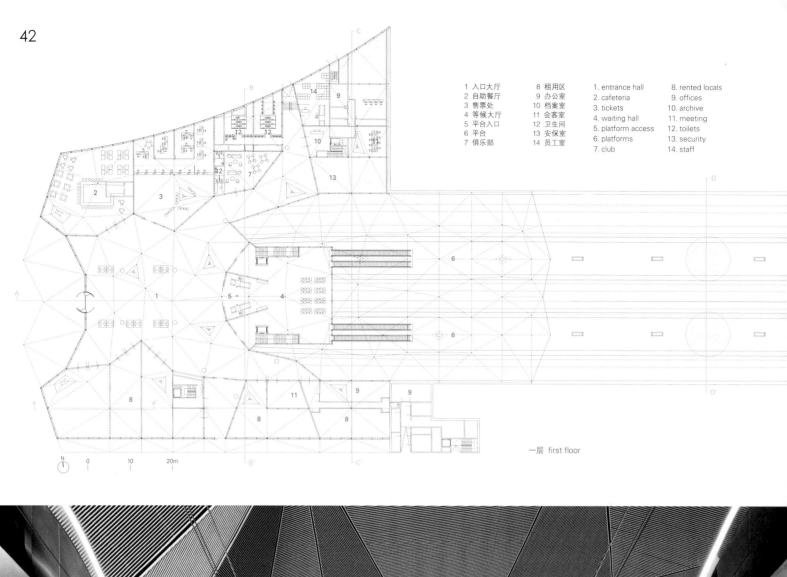

1 入口大厅	8 租用区	1. entrance hall	8. rented locals
2 自助餐厅	9 办公室	2. cafeteria	9. offices
3 售票处	10 档案室	3. tickets	10. archive
4 等候大厅	11 会客室	4. waiting hall	11. meeting
5 平台入口	12 卫生间	5. platform access	12. toilets
6 平台	13 安保室	6. platforms	13. security
7 俱乐部	14 员工室	7. club	14. staff

一层 first floor

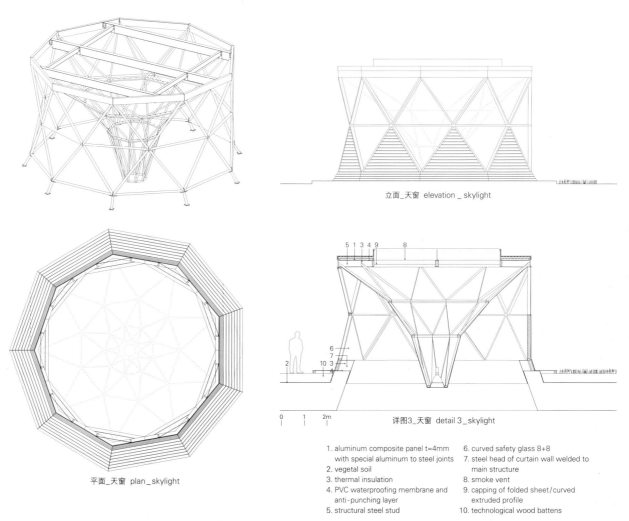

立面_天窗 elevation _ skylight

平面_天窗 plan _ skylight

详图3_天窗 detail 3 _ skylight

1. aluminum composite panel t=4mm with special aluminum to steel joints
2. vegetal soil
3. thermal insulation
4. PVC waterproofing membrane and anti-punching layer
5. structural steel stud
6. curved safety glass 8+8
7. steel head of curtain wall welded to main structure
8. smoke vent
9. capping of folded sheet / curved extruded profile
10. technological wood battens

米兰西北部地区的米兰展览中心酒店在2008年交付使用,而位于奇尼塞洛·巴尔萨莫中心区域,距离米兰只有几公里远的葛兰西广场于2004年开始重新设计,在这两项在意大利北部所取得的成就之后,建筑师多米尼克·佩罗在半岛南部的多个省份内都留下了自己的标记。得益于即将到来的地铁站项目,他对那不勒斯中央地带的主要公共区域——加里波第广场进行了彻底的改造。

这个广场纽带占地约六公顷,依傍于周围不规则的建筑文化遗产,至今仍被用作大型可用空间,来为联运和交通服务。这里没有合适的公共空间来举办城市生活活动,仅少部分空间供居民、游客和路人漫步和放松。

在未来,广场上即将出现一种崭新的交通基础设施,更突出了都市空间的多式联运特性,规模则与巴黎的共和国广场和巴士底广场加起来一样大。

几何学应用于地理

广场东侧设有具有历史意义的19世纪的广场和电车,确保与北部的翁贝托一世广场和南部的Nolana广场连通。

在广场北侧200多米长的距离内,一系列的公园、花园和引水点标记着通往地铁站的入口,为广场带来了新鲜元素。在景观设施的前方,一段大型开放式商业长廊嵌入了广场水平面下8m的位置,并渗入地下,通过一系列埋藏的画廊网格,加入到其他站点之中。这里是购物广场和鲜花市场。

在广场西侧中心站前,联运的露天广场成为火车站和汽车站(公交和出租车站)前的新滨海大道。车站的V形屋顶为高大的金属树冠结构,为新画廊遮荫。两座建筑结构在物理空间内较为接近,能够开展一段针对色彩和形式的辩证关系的对话。

纺织物制成的新屋顶以其坚定的现代感和非物质表现形式,书写了一段由建筑师设计的从站台到历史广场的三角关系模式。

地面的机械学原理

相反,在处于城市时代连接点的加里波第广场地下室中,却没有任何的考古遗址。这个布局设计处于工程初期,其建筑师提到:"事实上,

加里波第广场

Dominique Perrault Architecture

我们在某种程度上有点嫉妒其他那些有一点历史遗迹的地铁站,它们有一些历史遗留的痕迹。所以我们创造了我们自己的历史印记,并在地下嵌入一个大型画廊。原理就是地面之下的东西远远要比地面之上的东西更特别,或者至少是带有一些自然光。这个地下网络有时会出现,并将自然光引入深处,引入地面40m之下的地铁站。"

在广场中心区,自动扶梯不断地展开和折叠,在那不勒斯的天空下规律性地翻转和倒退。一座独立的、闪闪发光的、带有硬边的金属雕像的有规律的分布使人们忽视了那些为了确保基础设施结构的稳定而设立的令人印象深刻的柱子。

Garibaldi Square

After two achievements in the north of Italy – the NH-Fieramilano Hotel in the northwest of Milan delivered in 2008 and the redesigning of the Piazza Gramsci (2004), central place of Cinisello Balsamo, located a few kilometers from the Milan capital – Dominique Perrault makes his mark in the southern provinces of the peninsula, benefiting from the upcoming arrival of a metro station to orchestrate the in-depth-transformation of a major public space in the heart of Naples: The Piazza Garibaldi.

This square is a link of almost six hectares. It relies on a heterogeneous built heritage, and until now it is used as a huge available space serving the intermodality and the moving, without properly qualifying the public space to host the urban practices, and only offers to the residents, travelers and walkers few spaces dedicated to the ramble and the relaxation.

The future presence of a new transport infrastructure on the square underlines even more the multimodal character of an urban space, huge as the gathered Parisian places of la République and Bastille.

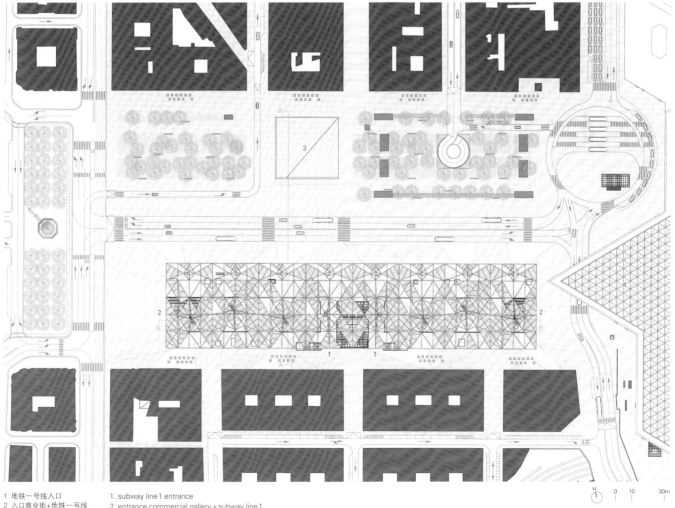

1 地铁一号线入口　　　　1. subway line 1 entrance
2 入口商业街+地铁一号线　2. entrance commercial gallery + subway line 1
3 地铁一号线+二号线入口　3. subway line 1 + line 2 entrance
4 那不勒斯中央火车站　　　4. Naples Central Train Station

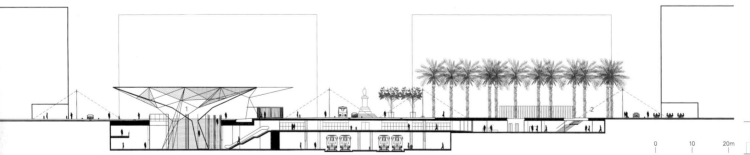

1 入口商业街+地铁一号线 2 地铁一号线+二号线入口
1. entrance commercial gallery + subway line 1 2. subway line 1 + line 2 entrance
A–A' 剖面图 section A-A'

The Geometry Serving the Geography

On the east side, there are the historic square of the 19th century and the tramway ensuring the connection to the north with the Piazza Prince Umberto and to the south with the Piazza Nolana. On the north side, on more than 200 meters long, a series of parks and gardens as well as a water point marking the access to the metro bring some freshness to the square. In front of this landscaping device, a huge and open commercial gallery, inserted into the ground at 8 meters below the square level, slides and seeps down under the ground to join, through a network of buried galleries, the other stations. Here are the shop square and the flower market.

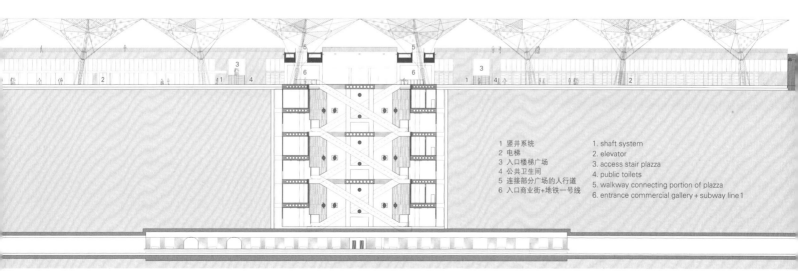

1 竖井系统	1. shaft system
2 电梯	2. elevator
3 入口楼梯广场	3. access stair plaza
4 公共卫生间	4. public toilets
5 连接部分广场的人行道	5. walkway connecting portion of plaza
6 入口商业街+地铁一号线	6. entrance commercial gallery + subway line 1

B-B' 剖面图 section B-B'

项目名称：Piazza Garibaldi / 地点：Naples, Italy
建筑师：Dominique Perrault Architecture
项目经理：Metropolitana di Napoli et Metropolitana de Milano
结构工程师：Nom Bollinger Bollinger + Grohmann / 甲方：Metropolitana di Napoli
功能：5 stations, garden, pedestrian area, commercial gallery, cinema, car park
用地面积：59,000m² / 总建筑面积：21,000m²
设计时间：2004 / 施工时间：2006 / 竣工时间：2014
摄影师：©Peppe Maisto (courtesy of the architect)

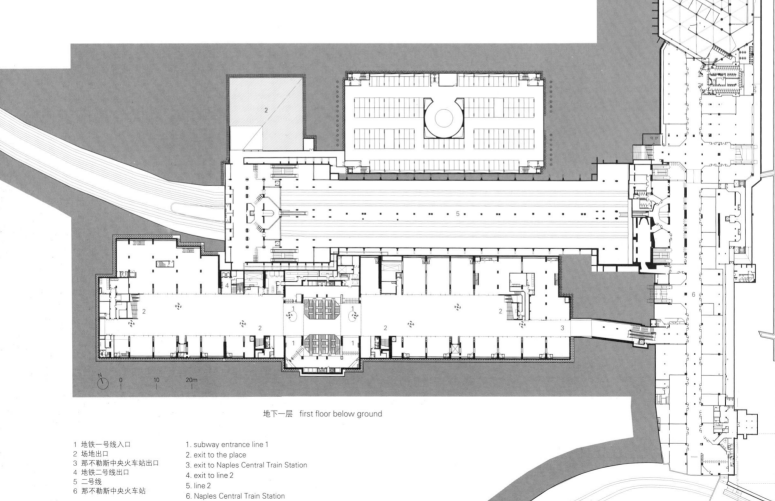

地下一层 first floor below ground

1 地铁一号线入口　　　　1. subway entrance line 1
2 场地出口　　　　　　　2. exit to the place
3 那不勒斯中央火车站出口　3. exit to Naples Central Train Station
4 地铁二号线出口　　　　4. exit to line 2
5 二号线　　　　　　　　5. line 2
6 那不勒斯中央火车站　　6. Naples Central Train Station

On the west side, in front on the central station, the intermodal piazza forms the new esplanade of the railway and bus stations (bus and taxi). The tops of the tall metal trees shading the new gallery are organized on the height of the V-shaped roof of the station. The physical proximity between the both structures opens a dialogue that is carried on in a chromatic and formal dialectic. Textile, the new roof extends in a firmly contemporary and immaterial form writing the triangular patterns designed by the architects from the station to the historic square.

The Mechanics of the Ground

With a touch of irony, the basement of the Garibaldi Square, located at the link of the ages of the city, at the inflection point of history, is devoid of any archaeological remain. This configuration is at the origin of the project led by the architect: "That is a fact, to some extent, we are a little jealous of the other metro sites, where there are some ruins, where there are some traces of history. So we created our own traces of history and we inserted into the ground a huge gallery. The principle is what is below the ground comes up in a special way above the ground, or at least, with some natural light. This underground network comes up sometimes and allows to introduce this quality of natural light into the depths, to the metro station located at 40 meters down into the ground."

In the central body, the mechanic escalators unfold and fold up. They regularly turn over and reverse under the Neapolitan sky. There is a single shiny metal sculpture with hard edges, whose regular arrangement ignores the presence of some impressive struts necessary for the structural stabilization of the infrastructures.

白湾邮轮码头
Johnson Pilton Walker

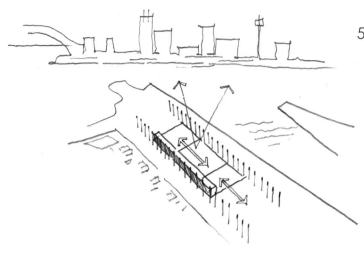

　　白湾邮轮码头是位于悉尼港的一个新邮轮设施,为澳大利亚迅速扩大的休闲邮轮产业服务。项目设计的突出特点是一个现代感十足的屋顶天篷,由具有历史意义的桥式起重机结构搭接起来。20世纪60年代后期,世界首个国际集装箱化航运服务在悉尼和欧洲国家之间开始运营,为纪念这里曾被用作船籍港,该结构被保留了下来。这种简单、抒情的姿态提供了一处灵活、明亮、宽敞的空间,适用于广泛用途,并为来到这座城市的游客们创造了一次标志性的、难忘的体验。作为这座城市的海洋遗产、特征和活动中有效的、活跃的、不可或缺的一部分,白湾邮轮码头代表了该地区发展的一个新篇章。

　　这个项目面临的挑战是要在这个退化的工业区旧址内为游客提供一段难忘和愉快的体验。设计并没有完全拆除现存的高架仓库,而是包容了场所的工业特性,保留了20世纪60年代大量的钢质桥架结构,将其作为主要的屋顶结构重新发挥价值。

　　通过剥离覆层、配套元素和冗余结构,北侧带有沧桑感的砂岩成为建筑的美丽背景,向东侧望去,美丽的城市天际线尽收眼底。设计对原有和重新利用的钢构件的整体性均进行了完整、细致的评估,建筑师并没有修补整个框架,仅对需要的部分进行了选择性的防腐保护和修缮,将原来的建筑痕迹留下来。通过新元素的植入,加之重新再利用的历史性肌理,该地丰富的海洋文化遗产清晰可见。

　　通过使用大规模工业用的桥式钢结构,单层亭子提供了一处大跨度的、灵活的空间。邮轮进港和出港的功能性需求设施分散地融入到建筑之中,方便了客流的流动,并且使船舶服务能够高效地进行。因为每年只有四个月的时间是邮轮使用旺季,而"没有船的"那些天这些空间就闲置了下来,用于举行各种公众活动,包括公共职能活动、展览、音乐会及招聘会。这种灵活的使用性使这处海滩在半个多世纪以来首次面向公众开放。

　　一处沿场地北部边界分布的砂岩悬崖已被列入了文化遗产,为航站提供了一个典型的悉尼风格的背景墙,作为场地在19世纪晚期创造的残留物,当许多海岬都被削减以增加海滨泊位时,这处悬崖设计增进了港口和临近的维多利亚时代的社区之间的关系。

　　波浪形的屋顶从头顶上方的新桁架和保留的起重机处开始悬挂,呈现一种表示欢迎的标志性姿态。这种形式是为了顺应大范围的工艺上和环境上的需求,其中包括内部日光照明、被动排烟口、雨水收集、声能吸纳,以及建筑内部与壮观的城市和港口视野的衔接。降低整体建筑的高度使港口周边设施的视野得以扩展。

　　悬崖一侧的室内不时地被一系列的白色箱子打断,回应了曾经堆放于此处的船运集装箱的形象和规模。这些简单的形态提供了员工便利设施和服务空间。箱子安置在柱子之间,跨过了景观设计内原有的、被刻意保留下来的铁路轨道,来为行人提供合理清晰的往来路线。

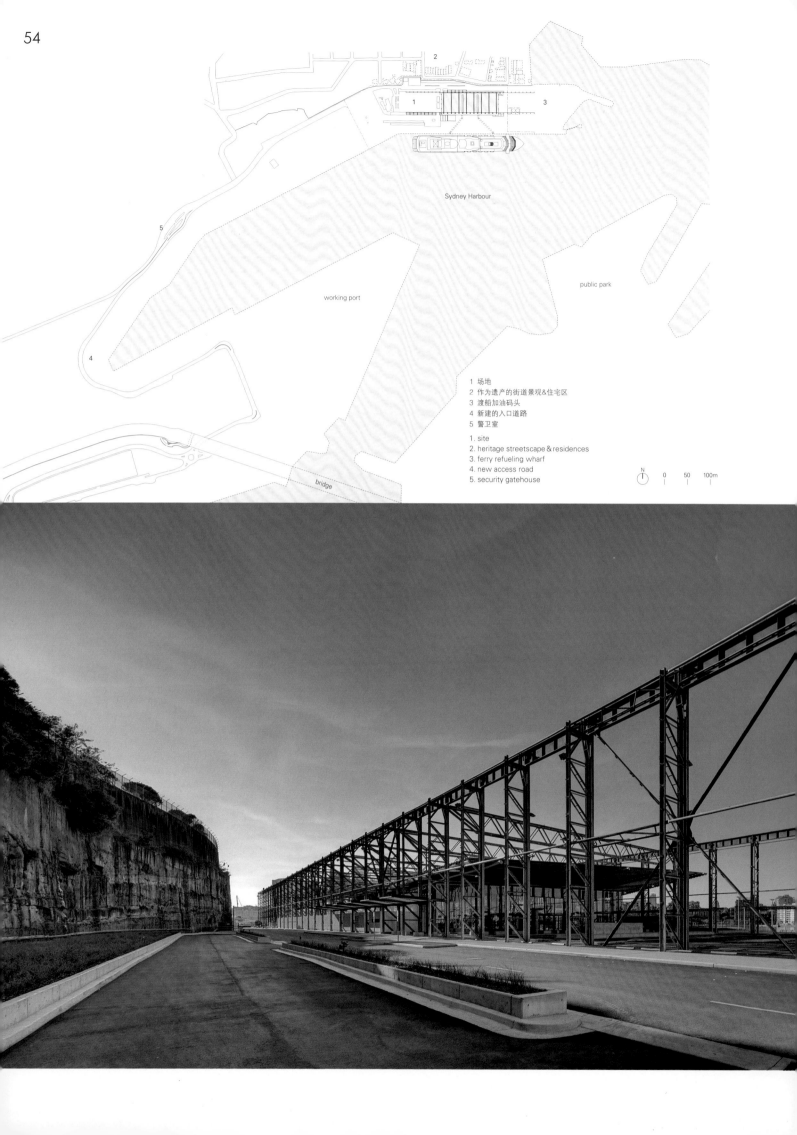

Sydney Harbour

working port

public park

bridge

1 场地
2 作为遗产的街道景观&住宅区
3 渡船加油码头
4 新建的入口道路
5 警卫室

1. site
2. heritage streetscape & residences
3. ferry refueling wharf
4. new access road
5. security gatehouse

White Bay Cruise Terminal

The White Bay Cruise Terminal is a new cruise facility located in Sydney Harbour, to serve Australia's rapidly expanding leisure cruise industry. The design features a contemporary roof canopy draped from a historically significant gantry crane structure. The structure has been retained as a memory of the site's previous use as a home port for the world's first regular international containerized shipping service, commenced operations between Sydney and European countries in the late 1960's. This simple, lyrical gesture provides a flexible, bright and airy space suitable for a wide range of uses, and creates an iconic and memorable arrival experience for visitors to the city. As an active, vibrant and integral part of the city's maritime heritage, identity and activity, the Cruise Terminal represents a new chapter in the evolving use of this place. The challenge was to provide a memorable and delightful experience for cruise passengers within a degraded industrial site. Rather than completely demolish the existing high-bay warehouse, the design embraces the site's industrial character, and retains the massive 1960's steel gantry structure as the primary roof structure for the new use.

1 前院
2 抵达大厅
3 行李领取厅
4 货舱
5 设备间
6 出租车停车区
7 长途公车停车区
8 行李保管库
9 观景露台
10 码头
11 景观
12 保安亭
13 商业厨房
14 商店
15 作为遗产的悬崖
16 作为遗产的铁路轨道
17 舷梯
18 游轮
19 停车场
20 报刊亭
21 流线

1. forecourt
2. arrival hall
3. baggage hall
4. cargo hall
5. amenities pods
6. taxis
7. coaches
8. baggage drop
9. viewing terrace
10. wharf
11. landscape
12. security booth
13. commercial kitchen
14. store
15. heritage cliff
16. heritage rail tracks
17. gangways
18. cruise ship
19. carpark
20. kiosk
21. pedestrian circulation

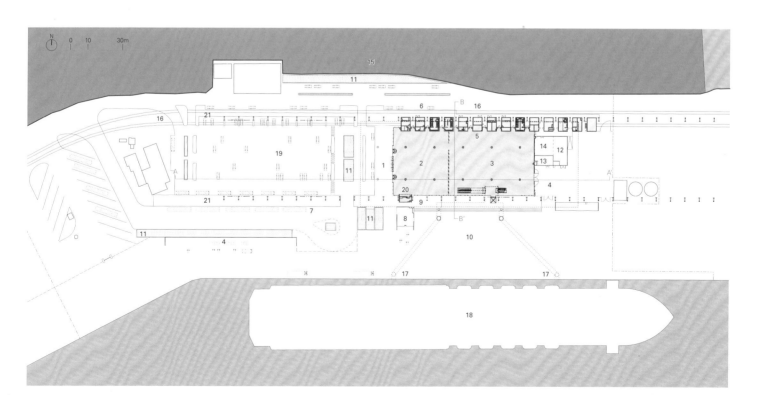

By stripping away cladding, services and redundant structural elements, the design reveals the dramatic, beautiful backdrop of the sandstone cutting to the north and opens up panoramic views back to the city skyline to the east. The structural integrity of the steel elements that were to be retained and reused was carefully assessed. Rather than refinish the entire framework, selective corrosion protection and repairs were applied where required only, leaving the scars of previous uses visible. By clearly articulating new elements alongside the re-used heritage fabric, the site's rich maritime heritage is made visible.

Utilizing the grand, industrial scale of the steel gantry, the single storey pavilion provides a large-span, flexible space. Functional requirements for cruise ship's arrivals and departures are discretely integrated into the building for efficient passenger movements and ship servicing operations. As the cruise season is active for only 4 months each year, on "non-ship" days the space can be emptied and used for a variety of public events including public functions, exhibitions, concerts and conference hire. This flexibility of use has enabled the foreshore site to be opened up to the public for the first time in more than half a century.

A heritage-listed sandstone escarpment along the site's northern boundary provides a quintessential Sydney backdrop for the ter-

minal. As a remnant of the site's creation in the late 1800s, when part of the headland was cut back to increase waterfront berths, the design's celebration of the escarpment reinforces the relationship between the harbor and the adjoining Victorian-era neighborhoods.

Hung from new overhead trusses and slung from the retained gantry, the undulating roof form is a welcoming and iconic gesture to the harbor. The form responds to a wide range of technical and environmental requirements including daylighting of the interior, passive smoke venting, rainwater harvesting, acoustic absorption and connecting the building's interior to wonderful city and harbor views. Views from surrounding properties of the harbor have been enhanced by lowering the overall building height. A series of white boxes punctuate the cliff-side interior, echoing the form and scale of the shipping containers once stacked on the site. These simple forms accommodate staff amenities and services spaces. The boxes are arranged between the columns, and over existing railway tracks that are retained in the landscape design to provide logical, legible pedestrian circulation.

1 拆除的工业棚的规模
2 作为遗产的钢结构
3 新屋顶
4 排烟口
5 古老的螺旋桨艺术品
6 作为遗产的后侧砂岩悬崖

1. size of demolished industrial shed
2. heritage steel structure
3. new roof
4. smoke vents
5. historic propellor artwork
6. heritage sandstone cliff behind

A-A' 剖面图 section A-A'

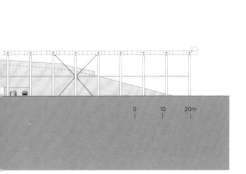

1 原有的结构	5 深水锚位	1. existing structure	6. cruise ship
2 新屋顶	6 游轮	2. new roof	7. heritage streetscape
3 原有的作为遗产的悬崖	7 作为遗产的街道景观	3. existing heritage cliff	8. existing caisson & industrial wharf
4 舷梯	8 原有的箱子&工业码头	4. gangways	
	9 原有的海岬线	5. deep water berth	9. line of pre-existing headland

B-B' 剖面图 section B-B'

1. existing bracing beam
2. top of pod parapet must align with glazing module
3. existing structural brace penetrates through roof sheeting, allow for flashing, colour to match roof sheet
4. structural brace in wall
5. 5 concrete topping slabs with saw-cut joints per external work plan
6. roof access walkway for maintenance to follow roof profile
7. 200x300 cutout in profile plan
8. suspended aramax aluminium gutter
9. stainless steel sump between existing stanchion column
10. penetration through column web for drainage
11. stainless steel downpipe to ground level draining to existing stormwater drainage with stainless steel bracketing
12. light-weight framing
13. 1x9mm layers of villaboard per acoustic engineer's requirement
14. 2x16mm layers of fyrchek per, per acoustic engineer's requirement
15. 75mm thick acoustic insulation per acoustic engineer's requirement
16. polymax thermal batt R3.0 roof insulation over 9mm fibre cement sheet
17. wall mounted split system per mechanical docuents
18. ceiling to meet noise reduction coefficient of 0.7 per acoustic engineer's report
19. plywood substrate
20. mechanical duct
21. 1x16mm fyrcheck
22. aluminum skirting colour to match wall cladding
23. polished concrete

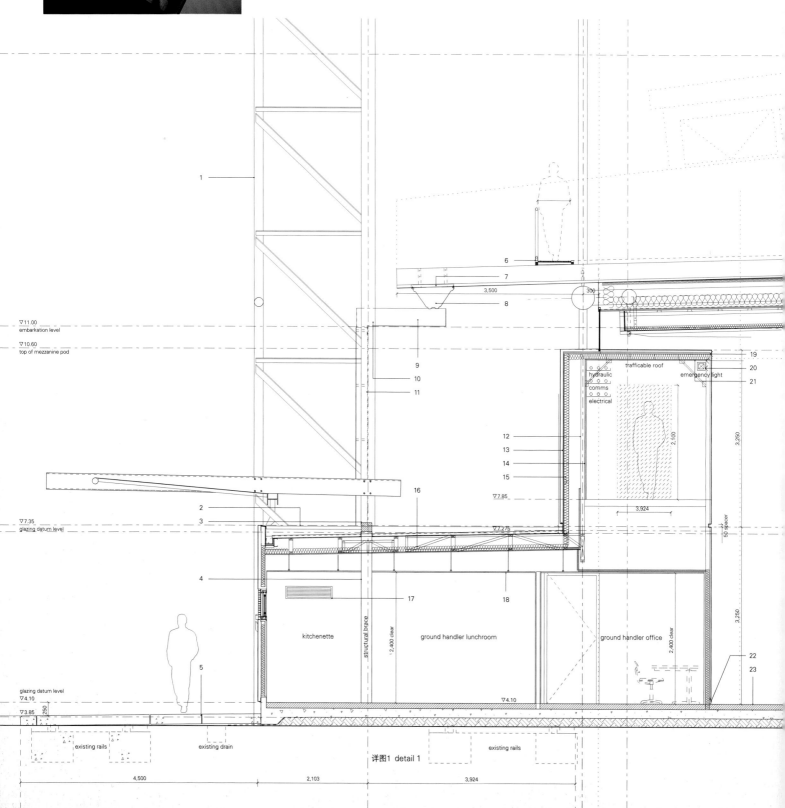

详图1 detail 1

项目名称：White Bay Cruise Terminal
地点：Sydney, Australia
建筑师：Johnson Pilton Walker
项目主管：Paul van Ratingen, Graeme Dix
项目建筑师：Brendan Murray
项目团队：Mathew Howard, Gareth Jenkins, Zoe Jenkins, Supinder Matharu, Natalie Minasian, James Polyhron, Chris Su, Daniel Upton, Andrew Christie, Adam Robilliard
结构工程师：Taylor Thomson Whitting
服务顾问：Hyder Consulting
消防工程师：Engineered Fire & Safety Solutions
承包商：AW Edwards
甲方：Sydney Ports Corporation & Barangaroo Delivery Authority
用地面积：16,000m² / 总建筑面积：9,600m²
造价：AUD 28m
施工时间：2012 / 竣工时间：2013
摄影师：©Brett Boardman(except as noted)

公共空间与城市规模项目
Public Spaces and Ur

公共空间和广场把人们、公众服务以及其他一些日常所必需的工作联系在一起。在研究这些建筑时，人们往往在周边选址、设计、便利设施等方面都给予了大量的投资。但是人们重点研究的还是这些城市连接区域成为当地社会和文化组成部分的经历以及不同的实现方式。购物街、广场均进行了升级、重新设计或重建，全新的竞争者也随着日新月异的需求如雨后春笋般出现。这些新的基础建设因通向或融入现有的景观，使其本身不仅把广场和商场联系起来，并且通过新的交集在现有城市的遗迹建立起连接。

城市需要公共空间和服务来满足居民和游客的需求。在城市、商场和广场内举行活动也是十分必要的。不论是从学校到河边的路程，或是拜访朋友的途中路径，或是商场内等候和休息的场地，还是欣赏远处的景观都是关键因素。人们停留的时间越来越少，但生活方式却越来越丰富。这就要求建筑和新的城市项目设计融入令人眼前一亮且创新的细节，来给人们提供刺激消费和社交体验的空间。

Public spaces and squares have become the connectors of people, services and other everyday essentials. Much is invested into surrounding location, design and amenities. Emphasis is placed on the experience and the different ways that these urban connectors have become social and cultural components in the public realm. Shopping malls and squares get updated, redesigned or rebuilt all together and new contenders spring up competing with the latest needs and services. With location for access and adding to the existing landscape, new infrastructure will enable the continued connections not just to the squares and malls, but also between the existing city heritage through the new intersections.

Cities need public spaces and services that fulfill the demanding requirements of its residents and visitors. Essential is also the in-between activity that takes place in the city, the mall and the square. The walk between the school and the river, the journey to meet friends and the place to sit and wait in the mall whilst admiring the landscape beyond are all key factors. With less time and more demanding lifestyles, the architecture and design of new urban projects incorporate exciting and innovative details, providing us with stimulating retail and social experiences.

Pormetxeta广场_Pormetxeta Square/MTM Arquitectos + Xpiral Arquitectura
辛特拉论坛_Forum Sintra/ARX
曼福综合馆_Mapfre Complex/TSM Asociados

公共空间与城市规模项目_Public Spaces and Urban Scale Projects/Heidi Saarinen

ial
ban Scale Projects

"如果对街道多样性的全天应用能够极大地丰富消费者的需求和品位,那么所有独特的都市专业服务和商店都能应付自如,这是一个以自身为基础的过程。"简·雅各布[1]

随着当今社会经济的高速发展,越来越多的灵感和确切的需求促使城市区域在时间上和空间上连接,以此来有效地解决在地理范围内就可以满足人们的购物、社交、文化和交流需求。

很多地方都迅速地被新的城市广场和购物区域或商场改造,这种改造通常有着混合型用途,且设在战略性位置上。我们对合适的设施的需求越来越多,以使它们能够更迅速地适应购物疗法或日常琐事的统一性——在去咖啡休息室会见客人的路上拜访一下药剂师,然后在夜晚来临时谈成一笔生意。通过明智的设计方案和合理的空间利用,所有的需求都能够在一处区域得到满足。

那到底是什么打动了消费者、游客和路人的心呢?对于设计过程和设计可行性而言,满足特定区域以至国家的需求的能力,以及兼顾气候、途径以及地点的考虑是至关重要的。人们越来越期待更灵活的零售店和其他批发商。合理的室内利用、更多的服务、密集的客流量以及华丽的建筑外观会提高用户和游客的感受和满足感。

本文对三处不同的场地和建筑进行了调查,它们都有自己独特的特点和用途,分别是ARX设计的位于葡萄牙IC-19大道上的辛特拉论坛、TSM协会设计的位于秘鲁独立广场的曼福综合馆,以及由MTM建筑事务所和Xpiral建筑事务所设计的西班牙巴拉卡度的Pormetxeta广场。

公共空间和城市规模项目

ARX设计的这座恢宏的建筑的场地面积为55 000m², 是为辛特拉

"If this diversity of street use spreads a variety of consumer needs or tastes throughout time of day, all sorts of uniquely urban and specialized services and shops can make out, and this is a process that builds upon itself". Jane Jacobs[1]

With today's fast paced socio-economic climates, there are ever-increasing aspirations and genuine needs for spatio-temporal connecting points in the urban realm; to fulfil a host of desirables; retail, social, cultural and communicative needs effectively, aesthetically and within geographical reach.

Many areas are quickly transformed by new city squares and shopping areas or malls, often with mixed uses and placed in strategic locations. More and more do we require adequate facilities for the quick fix of retail therapy or oneness with the daily chores – popping into the chemist on the way to the meeting in the coffee lounge, then closing a business deal before venturing out for the evening. All sorts of needs can be catered for in one location, through clever design strategies and use of space.

What is it then, that sets the pulse for the consumer, visitor and passer-by? Ability to customize specific geographical and national needs, and taking into account climate, access and location are important factors for the feasibility and design process. Demand for more flexible retail and other commercial outlets is growing and simply expected. Smart use of space, wide range of offered services, high visitor footfall and architectural panache weigh up the consumer and visitor experience and satisfaction.

In this article, three different sites and buildings were investigated, all with unique characteristics and uses. The buildings/sites covered are Forum Sintra on the IC-19 road in Portugal by ARX, Mapfre Complex in Independencia, Peru by TSM Asociados and Pormetxeta Square in Barakaldo, Spain by MTM Arquitectos and Xpiral Arquitectura.

Public Spaces and Urban Scale Projects

The empowering structure on the 55,000m² site for the development of the Forum Sintra shopping centre and square by ARX is landscaped into the IC-19 road system and site of a demolished former shopping mall and cinemas. The design allows for a new, indoor shopping district echoing the dramatic topography of the

论坛购物中心和广场的开发而设计的,将景观融入IC大道体系以及拥有一个废弃的购物街和电影院的场地中。设计形成了一处新的室内购物区域,它呼应了佩纳宫下方的辛特拉山周围环境的险要地势。建筑师这样描述道:这一体块被分成两个垂直且相对的部分,一个是厚重的黑色混凝土基底,把建筑和地面连接起来;而另外一个则向后,饰面为白色,是典型的葡萄牙传统建筑。外观、形式和规模与内在的社会经济价值的特征,以及百搭的主题把景观和购物结合,作为一个事件,一个必需品,一个社会和文化的熔炉,这些都是当代的核心内容。

外部绵延耸立的立面和长长的水平面在前面延展,且在某些地方直冲云霄。混凝土和钢筋的清晰线条和粉刷的表面在某些交点对齐,形成这个复杂设计的统一性。景观园林将中间的空间、入口和出口路线连接起来,以构成更大的框架和使用空间。

独立的零售单元非常有特色,排列在内部购物街的两边,形成辛特拉论坛最主要的流线。在路的外侧,类似于开放空间、咖啡厅和餐厅的广场区延伸进开放的中心空间,连接着其他地区的景观,似乎要邀请其入内。这个综合体的内部设有指定的概念区,如"绿色商场",代表了精致的"绿色玻璃层";以及"家庭式商场",内部为木质层压饰面,暗含了温馨的居住环境如家庭般温暖。这些主题来源于材料和需求的并置和对比,构成文化和美学氛围,或者小型城中城(连接处设有广场)。自然光透过零散且不规则的天窗照射进来,像宝石一样。玻璃、钢、陶瓷、铝质品、石头与带有格局的、间隔出现的木质天花板相互作用。这些设计都是为了在特定的主题区域和公共室内小径中营造个性独特的用户体验。

引导指示牌和图像信息牌环绕整座建筑,沿着建筑的轮廓以大型

swaying surroundings of the Sintra mountains with the Pena Palace on its top. *"Its mass is divided in two opposite vertical extracts: a heavy base in black concrete, connecting the building to the ground, and a second volume, pushed back, in white finish, as a clear reference to Portuguese tradition"*, describe the architects. Identities of shape, form and scale with contained socio-economic values, and its themed "suit-all" stance, connect the landscape with the use of shopping as event, necessity, social- and cultural melting pot, all central to contemporary existence.

The exterior's staggered and rising elevations and elongated horizontal planes extend along the front and in places appear to "take off" towards the sky. Clean lines of concrete, steel and rendered surfaces align at many intersections to form the unity that this complex design achieves. Landscaped gardens link the in-between spaces, entrances and exit routes to form a larger framework and use of space.

Individual retail units with particular identities array along the interior mall paths, culminating into the main Forum Sintra's circulation. Out of the arrays, plazas like open spaces, cafes and food hall extend into an open central space, connecting and inviting in the landscape beyond. Within the complex, designated concept areas, such as the "green mall" representing the delicate green glass floor and the "home mall" with wood laminated interior applications, hinting at warm dwelling space associated with home. These themes were derived from the juxtaposition and contrast of materials and needs, creating cultural and aesthetic atmospheres, or a mini city within a city, with squares at the junctions. Throughout, natural light enters the space through scattered irregular skylights, simulating gems. Glass, steel, ceramic, aluminium and stone interplay with patterns and occasional timber ceilings, all carefully planned for the purpose of creating individual and special experiences in the designated themed areas and public indoor paths.

The signage and graphic communication wrap around, cut into and morph along the silhouette of the building in form of a large scale metal sign, setting the scene from a distance and can be used as a mapping locator and advertising from far away.

The Mapfre multi-functional and multi-level commercial complex in Peru is organised in blocks of mixed-use spaces in a contem-

巴拉卡度的Pormetxeta广场，比斯开省，西班牙
Pormetxeta Square in Barakaldo, Vizcaya, Spain

金属标志的形象变幻着，使其对轮廓进行了分割，使人们在远处就能看见这处购物中心，并且成为独特的景象，可以用作地图定位器和远处显眼的广告牌。

在一系列的当代配套建筑中，秘鲁的多功能、多层次的曼福综合馆设有分割的混合使用空间，其旁边是曼福保险公司和教堂。该设计还包括一个二层的医疗中心和一个六层的骨灰安置所，人们可以从高耸的二楼俯瞰入口左边的可容纳100人的教堂。

建筑平面内，其他相连的空间，如用于守灵和集会的区域，安排在主广场的四周。通道在保险公司和教堂之间，便于公共空间和私人空间的便利和分化。

椭圆形的遮阳板遮挡着骨灰安置所的洞口，保护室内的隐私和进行的活动，同时，一系列从光亮透明到黑暗模糊的不同类型的玻璃板用作外立面，让光进到入口右边的医疗中心、办公大楼和远处的人行道。

朴素的特性显著地体现在教堂内部，黑色大理石材质的圣坛以低调的形式高度实现了整体的一致性。室内和室外部分都覆盖了从秘鲁阿雷基帕州开采的石板。薄木板和教堂前后的石墙形成对比，呼应着简单而优雅的木质长椅。从上面天窗自然照下的亮光，以及添加的LED照明，映衬了祭坛的木质背景，似乎保护着此处空间这一时刻的宁静。教堂被设计成中立的朝拜圣地，提供一系列的关于信仰的服务。教堂和骨灰安置所通向守灵区的入口，旁边就是人行道，人们借此道可以进入公共区域。在这里，规划的私人活动的间歇时间可以进行集会和非正式的活动。专用电梯可以到达这一层和地下室的太平间。

还有一个进出的通道，即沿着通往楼上保险公司行政区和广场的开放楼梯和坡道，那里是另一个交会连接点。就建筑群的规划、进程以及

porary co-ordinated set of buildings where the Mapfre Insurance Company and a chapel occupy neighbouring spaces. The design further comprises a two-storey medical centre and a six-storey columbarium, towering from the second floor, above the chapel, seating 100 people, located to the left of the entrance.

Other linked spaces, such as areas to carry out wakes and gatherings are situated around a main square on the building plan. Walkways are assigned between the insurance block and chapel unit, for easy access and differentiation between the public and private functions.

Ellipsoid sun louvers cover the openings of the columbarium, protecting the privacy and activity of the interior, whilst a range of different types of glass panels stretching from "light and transparent to dark and almost opaque", are used as facades, letting light into the medical centre, located to the right of the entrance, office blocks and walkways beyond.

Notable simplicity is seen in the chape's I interior, where the humble form of the dark marble altar justifies the holistic purpose beautifully. Flagstone, quarried from the Peruvian state of Arequipa, clads parts of the interior and exterior. Wood veneers contrast the stone to the front and rear of the chapel, echoing the simple yet elegant wooden benches. Naturally lit from a skylight above and with added LED lighting features, the wooden backdrop of the altar appears to protect the tranquility of the moment in this space. Designed as a "neutral temple", it welcomes a range of faith services.

A communal area can be accessed via a walkway, which is next to the chapel and columbarium leading towards the entrance of the wake area. Here gatherings and informal interactions can take place between private scheduled events. There is a special lift access between this level and the basement mortuary.

Another access point is the second walkway, along the open staircase and ramp leading to the upper level of the insurance administrative block and plaza, another meeting and connecting point. Materials, lighting and circulation have been coordinated respectfully throughout, considering the programme and events alongside the commercial aspects housed within the complex. Particularly elegant are the access routes, a kind of memorial journey,

及商业因素而言,材料、照明和流线始终保持协调。作为纪念之旅的砂岩质地的通道可谓优雅之至,此外教堂内的表面均由原木装修,也是优雅的典范。

另一方面,在毕尔巴鄂附近巴拉卡度的老工业区里,有一个类似Pormetxeta广场的飞机库,它以耐心细致的态度,随意地使用了各种材料,利用不同的高度和远景来招揽游客。广场的表面设计采用了特定的模式,使材料在节点处交汇,且随意发散出去,此外又加上全新的颜色、光线和纹理,以此来引导游客的观光之旅,延伸至尽头。多孔钢网屋顶和悬臂遮篷条俯瞰着四周的环境,似乎在与相通又相离的现代房屋开发和建筑遗产捉迷藏,娱乐感十分突出。六角形拼凑式的坡道带领游客、行人和乘客穿过当地的各个场所。

穿过广场时,路人可以驻足停顿,在凉亭和圆顶地下室内遮风避雨,并欣赏游乐区和城市丛林区(即广场)。这里没有自然植被,取而代之的是纵横交错的人造颜色、材料和投影。中央天篷结构的顶部,或者被岩石网覆盖的钢形树架会招来一些野生动物栖息于此。

这个地方建于废弃空间和老工业园区之间,增建的人行道、起拱面和混凝土基底之上的街道设施,加之一些极具创新的材料形成了一个集合点、连接器和城市的门槛,在诺温河、铁路和城市与河流之间的过道的影子映衬下,把这座城市的其他街道网格引领到此,达到极致。因为之前的设计没有有效地连接和服务社区,因此建筑师试图从先前的城市总平面中创建一个具有新建入口和连接点的"地理特征"。

城市、广场和河流相连接的地方呈开放状态。学生上学或短途旅行时穿过广场,欢快跳跃,享受这富有想象力的景观。

建筑师的图纸流露出实际的视觉感受。夜间,广场变换为光与长影

with the texture of the sandstone, and timber clad interior details within the chapel.

On the other hand, the hangar like Pormetxeta Square, located on a former industrial site in Bakaldo, near Bilbao has a no-fuss attitude, greeting the visitor with a playful mixed use of materials, heights and vistas. The surfaces of the public square are designed in ad-hoc patterns, materials meeting at junctions and then taking off in new directions, adding colour, light and texture, guiding the journey of the visitor, extending to the vanishing point.

Steel mesh perforated roofs and strips of overhanging canopies, catch glimpses of the surrounding neighbourhoods, giving the impression of teasing collaboration of hide and seek with the existing housing developments and architectural heritage that share and border this site. Hexagonal patchwork paved ramps take visitors, passers-by and commuters across the site, acting as a through route between local destinations.

The route through the square allows passers-by to stop and pause, take shelter under the pergola and undercrofts and enjoy the games zones and urban jungle that is this square. Natural vegetation has been replaced by applied manmade colour, materials, shadow play and criss cross forms. The tops of the central canopy structures, or steel trees are covered in nets of boulders and may invite some wildlife habitats to form.

Place has been created here, in what was an in-between, neglected space and previous industrial estate. The addition of the paved walk ways, street furniture on springs and concrete bases together with the curiously innovative use of materials throughout now act as a meeting point, connector and city threshold, culminating from the rest of the city's street grid into this site, in the shadow of the Nervion River, railway and throughways between the city and the river. The vision of the architects was to create a new "geographic identity" from a previous urban site plan that did not connect or serve the community effectively, with recreated access and new uniting points.

The connections of places between the city, the square and the river have been opened up. School children cross the square on way to school and excursion, enjoying skipping along this imaginative landscape.

投射下的迷人景观,灯光闪烁的空间激起了人们的好奇心,又不失安全感。建筑师解释说:"我们将一个城市战略设计为一个公民之间的商业合作和对话的平台,来实现新型的社会凝聚力。

商场(和公共空间)不仅成为城市的基本建筑群,也成为城市连接、访问和结合的最好的工具之一。"[2]

总之,人们需要购物和体验其他必要并令人向往的任务,日常生活因此变得快捷而有效,愉快而难忘,而其中奢侈的便利设施也在明显改变。我们的行动更快捷,决策更坚决,信奉"时间就是金钱"的生活模式。在一个屋顶,或者一个购物中心和广场内就能满足多种需求和服务的建筑需求才是人们遵循的范式。

例如,河流和市中心的景观、道路、交通网、生活设施和地标都是整个设计的关键。我们有时可以看见在城市外围扩建的购物区和大片区域,在适当的位置设有道路和精细的基础设施,方便车辆进出和停泊。对空间、服务、形式和材料的使用正在发生变化,也能看到一些尝试,新要求和创新,如Pormetxeta广场和曼福综合馆,建筑被混合使用且形式不断变化,在本章介绍的所有项目中,特别是辛特拉论坛,多角度的延伸仿佛在邀请和吸引着顾客的进入。

此外,作为城市的连接器、节点和目的地,商场和广场的重要性变得非常明显,且体现在本文介绍的所有场所中。传统购物中心的服务不能完全符合指定的社区以及发展的客户和用户群,显然,现在有更高的需求。

1. J. Jacobs, *The Death and Life of Great American Cities*, London: Pimlico Press, 1961(2000 edition), p.174.
2. C.J. Chung, J. Inaba, R. Koolhaas, S.T. Leong, *The Harvard Design School Guide to Shopping*, Harvard Design School Project on the City 2, Taschen, 2002, p477.

Architects' drawings radiate the feeling of the actual vision. Nighttime, this square transforms into a dramatic landscape with elongated shadows and lights, with illuminated spaces creating intrigue and sense of safety. The architects explain: *"Consequently we have planned an urban strategy as a business and dialogue platform between citizens, creating a new social cohesion".*
"Not only has shopping center (and public spaces) become the basic building block of the city, it has, moreover, become one of the best tools for providing urban connectivity, accessibility and cohesion". [2]

To conclude, the need for shopping and other necessary and desirable tasks, makes the daily routine speedy and efficient, enjoyable and memorable, while more extravagant amenities are clearly changing. We move faster, make quicker decisions and "time is money" is the pattern of today. The need for buildings that can accommodate a host of needs and services all under one roof, or rather, in one mall or square, is indeed the norm.
Landscape, road and commuter networks, local amenities and landmarks, such as rivers and city centres are key to the overall plan. In some instances, we see spreading shopping estates and large areas outside of the city, with roads and sophisticated infrastructure in place for easy access and parking facilities. Use of space, services, form and materials are evolving and some experimentation's and new demands and innovations are seen, such as in the Pormetxeta Square and Mapfre Complex, in the mix of uses and the changing form as is seen in all of the buildings here and particularly in the Forum Sintra, extending out into the site at all angles, inviting and enticing the consumers in.
Moreover, the importance of malls and squares as urban connectors, nodes and destinations has become apparent, and can be seen in all of the sites covered here. The services of the traditional shopping mall no longer fully serve the community of demanding and evolving customers and user-groups, clearly, there is now a high demand for more. Heidi Saarinen

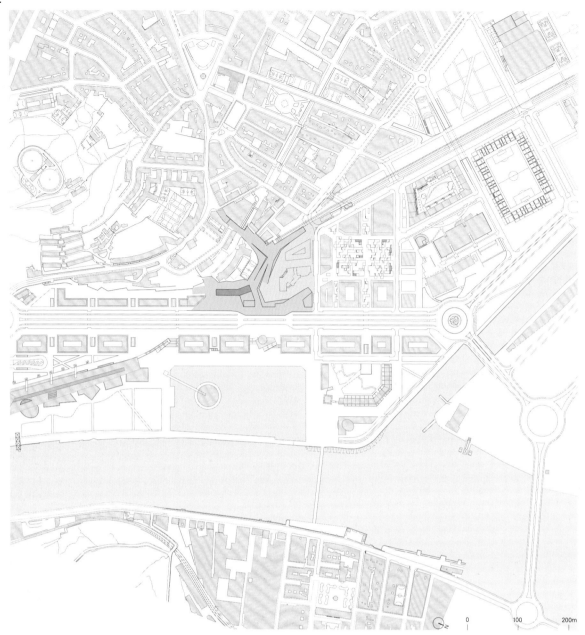

巴拉卡度Pormetxeta广场

MTM Arquitectos + Xpiral Arquitectura

工业属性

巴拉卡度区位于诺温河上一处非常陡峭的斜坡场地内，在20世纪60年代，这里曾经开发了一个综合城市规划，其中比斯开高炉成为主要的建筑。建筑师提出为其赋予一个全新的地理特征，忘掉Urban-Galindo曾经为这片区域设计的城市规划，并且将其移出这片区域。

城市渗透性的实质

建筑师对原有的规划进行了调整，以建造全新的巴拉卡度广场（25 223m^2），旨在为城市创造全新的优势：一方面它是通往城市中心的一个全新的入口点，另一方面它提供了城市与河流之间更好的流畅性和移动性，这是因为在城市的渗透性规划中，工厂被关闭了。

配备齐全的交通干线

场地内的非常陡峭的斜坡已经成为创造一个全新的城市类型的机遇。利用原有的水平层（高处与低处之间的水平差有20m）之间的差异，建筑师建造了一个配备齐全的斜坡，即Pormetxeta广场，策略如下：

首先，所有的入口和道路都进行了连接，包括铁路站台、高速公路、面向河岸滨道的入口、人行道和街道、主街道入口、地势低的街道区域、沙漠广场以及学校。

其次，建筑师将Pormetxeta广场斜坡下面的所有空间进行了整合，以建造2341m^2的商业和服务性公共设施。

第三个优势是将广场作为一个微公共空间，包括微广场、会客区、休息站台、室外逗留区，换句话说，要按照一定的序列产生一系列的现象。

最后一个优势是广场将作为一个多功能元素，即街道设施、栏杆、凉亭、灯笼等等，以形成一处配备齐全的地方。

充满生气的广场

Pormetxeta广场位于长长的斜坡之下。这里有一些孔洞，用来限定设有游乐场的空间的范围，并使其成为一处独特的地形。几棵大型人工树高11.5m，由考顿钢材制成，树梢处放置了卵石。它们将作为遮阳伞，下面充满了儿童、溜冰者的笑声以及谈话。由于其间的差异性，Pormetxeta

照片提供：©Suravia Fotografia (courtesy of the architect)

广场和不同的水平路线既构成了一个独特的项目，同时也是一个独立的项目，并且还突出了其用途。

建筑与都市生活

建筑师的目标是建造一座城市，将公共空间和私人空间理解为城市规划和教化模型的混合体，并且建筑与都市生活恒久地交织于此。因此，建筑师制定了一个城市战略，来作为市民之间的商务对话平台，以形成一个全新的社会凝聚力。

Pormetxeta Square in Barakaldo

Industrial Identity

The district of Barakaldo is located in a very steep slope territory above the Nervion River, where a complex urban planning was developed in the Sixties and was leaded by "Blast Furnace from Vizcaya". The architects have proposed to provide it with a new geography identity, by forgetting and removing the last urban planning that Urban-Galindo had created for this area.

Marrow of Urban Permeability

The architects modified the existing planning in order to create the New Square of Barakaldo (25,223m²), for the purpose to create new advantages for the city: on the one hand a new access point to the urban center, on the other hand a better fluidity and mobility from the city to the river, as the industrial factories were shut in the permeability of the city.

Equipped Communications Arteries

The very steep slope of the plot has become an opportunity of creation of a new urban typology. Taking advantage of the height

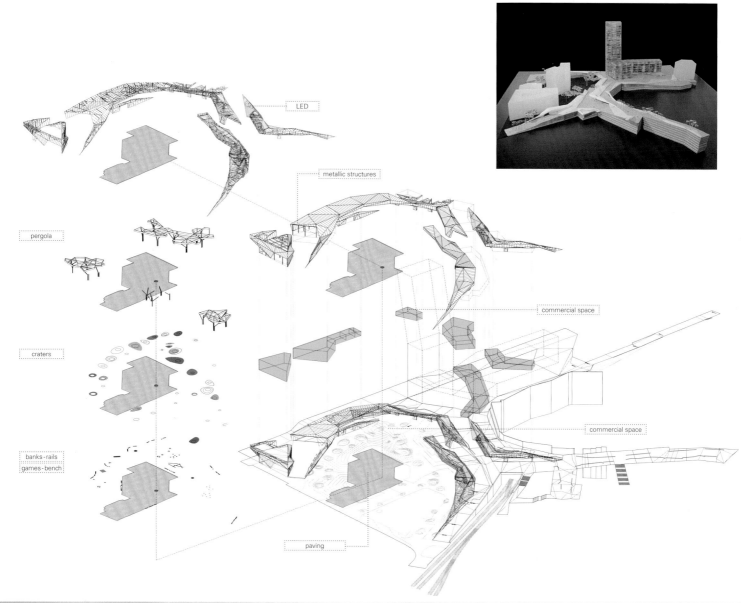

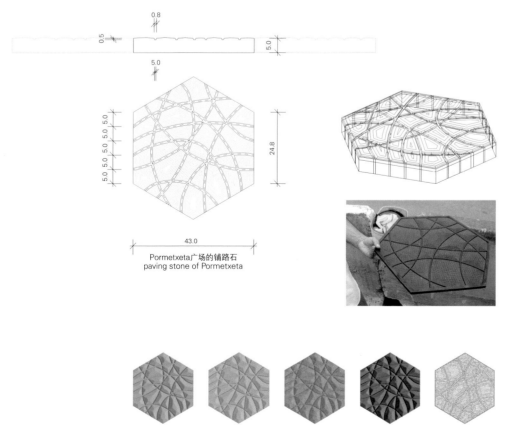

Pormetxeta广场的铺路石
paving stone of Pormetxeta

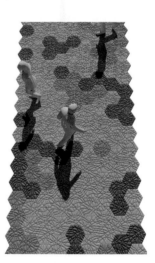

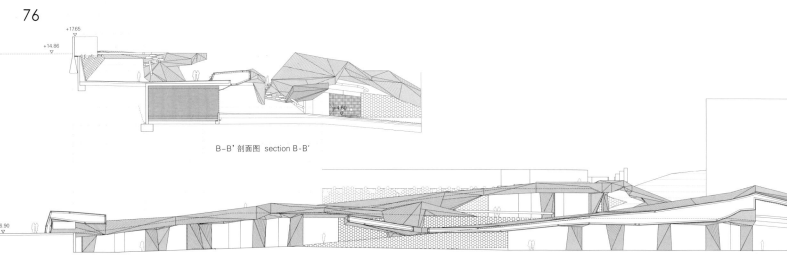

B-B' 剖面图 section B-B'

A-A' 剖面图 section A-A'

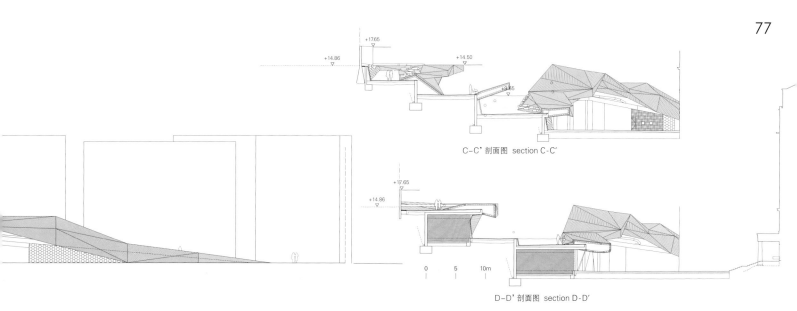

C-C' 剖面图 section C-C'

D-D' 剖面图 section D-D'

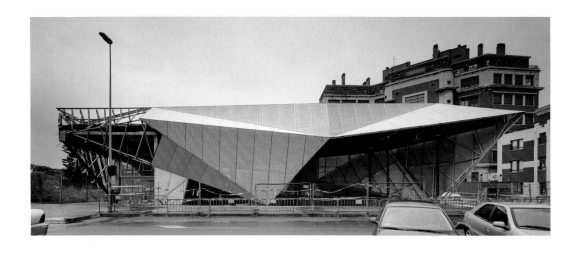

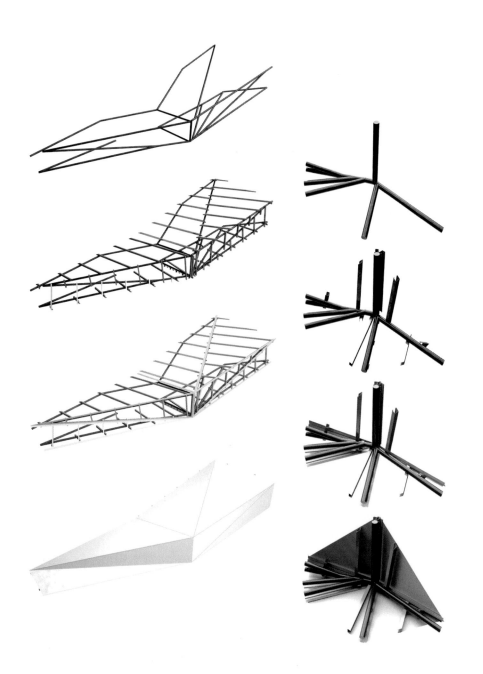

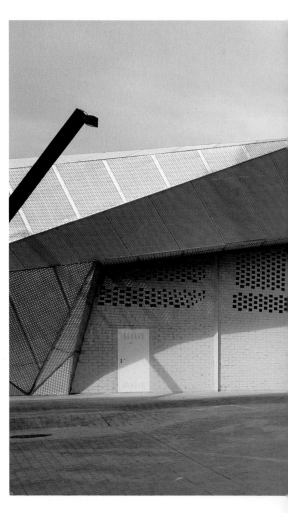

项目名称：Pormetxeta Square, in Barakaldo (Bilbao)
地点：Barakaldo, Vizcaya, Spain
建筑师：MTM Arquitectos_Javier Fresneda Puerto, Javier Sanjuán Calle / Xpiral Arquitectura_Javier Peña Galiano
合作者：MTM Arquitectos_Jesús Barranco, Álvaro Maestro, Miguel García-Redondo, Zaloa Mayor, Marianne Richardot, María Rey, Fátima López, Álvaro Maestro / Xpiral Arquitectura_Malte Eglinger, Sergio Corredor, Maren Kläschen, Maud Thiery, Daniel Cano, Fernando Such, Lola Jiménez, María José Marcos
照明工程师：Mario Gentile
结构工程师：I.D.E.E.E. Eduardo Diez Gernier
Systems Building Grupo JG. Amador Muñoz
建筑工程师：Jose Miguel Ortega, Igor Ortega
技术助理：Dair ingenieros, Jose Ángel Laguna, Enrique Álvaro
施工单位：U.T.E. Pormetexeta (Sacyr + Pabisa)
发起人：Bilbao Ría 2000
建成面积：residential_10,856m², parking 4,049m², commercial 2,341m²
造价：EUR 8,543,353.44
竣工时间：2010
摄影师：©David Frutos (courtesy of the architect)(except as noted)

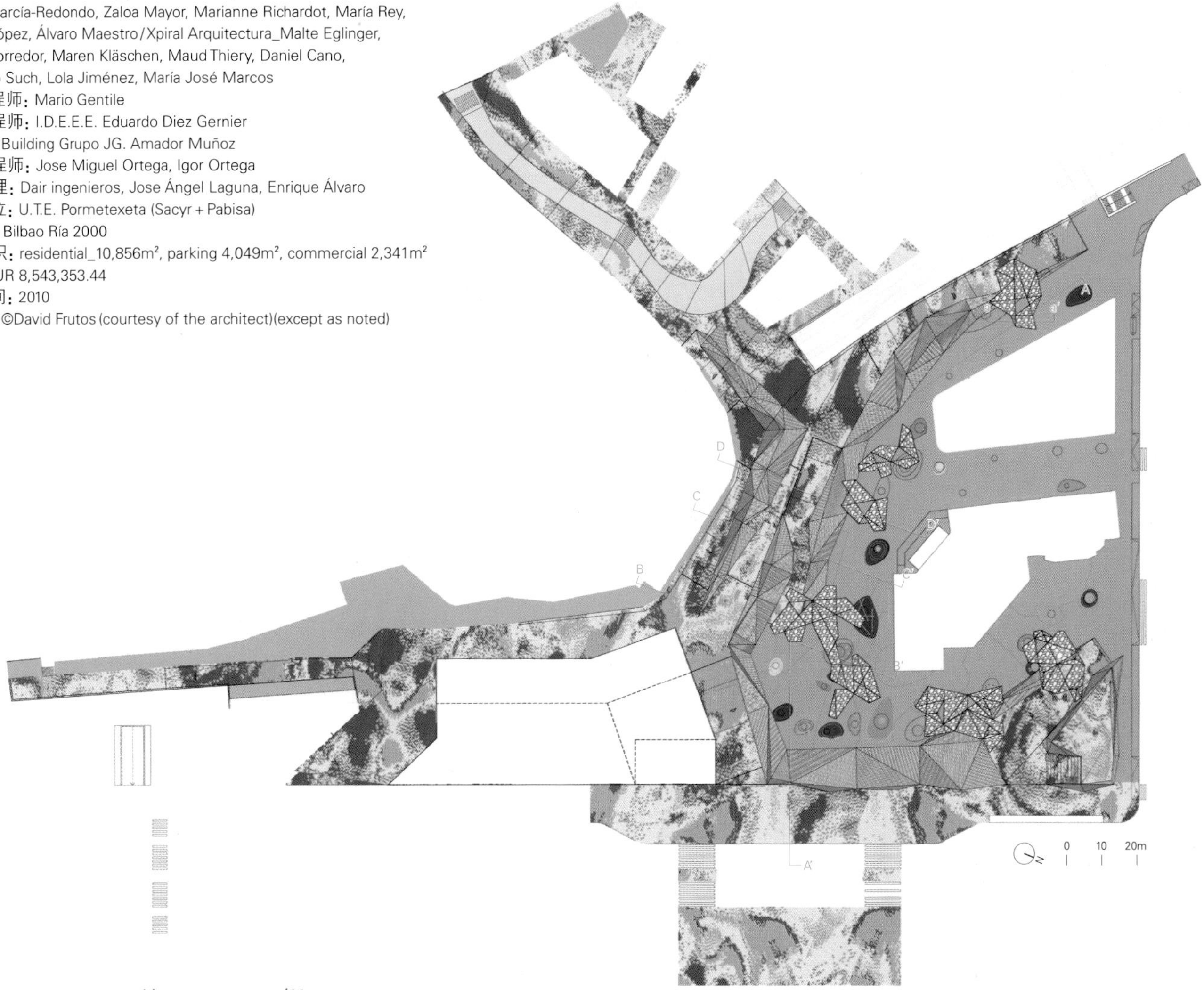

二层 second floor

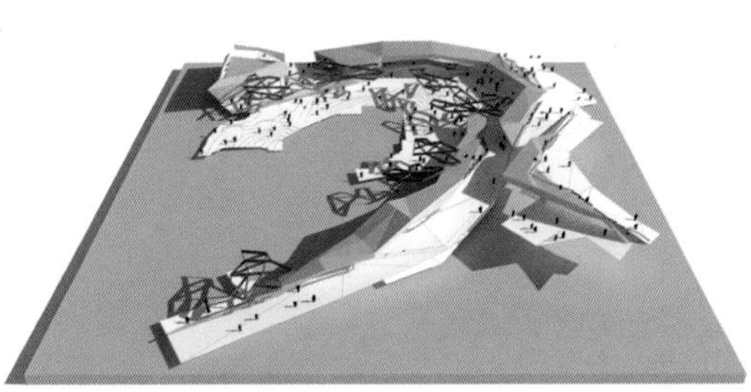

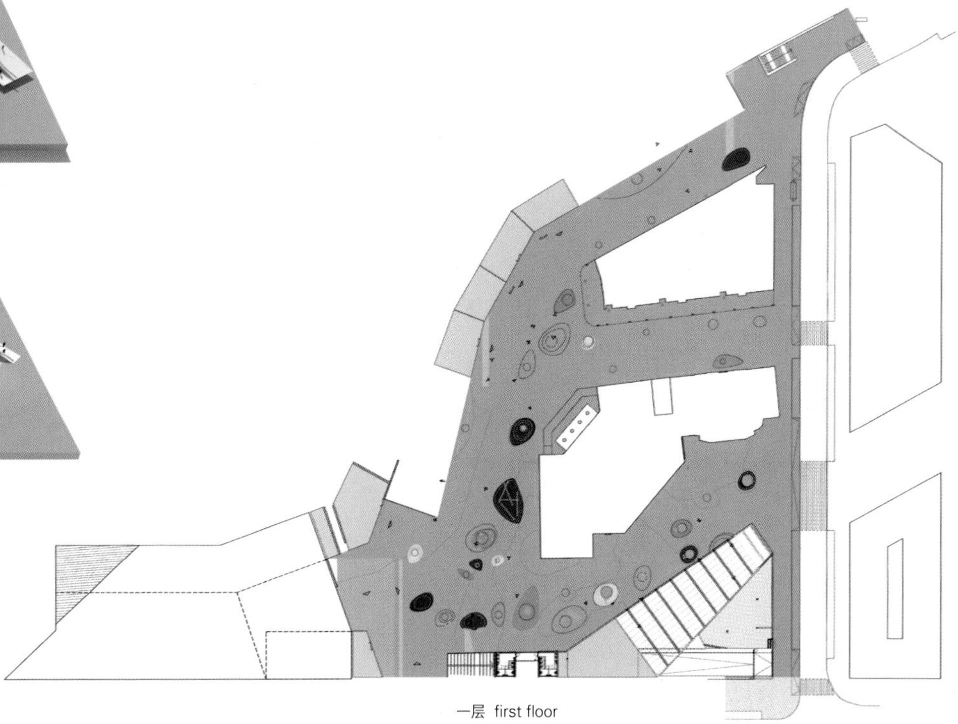

一层 first floor

difference between existing levels that is 20 meters between the upper level and the bottom level of the plot, the architects built an equipped slope, "the Pormetxeta Square", with the next strategies:

First of all several accesses and approaches are linked: the railway platform, the highway, the entry towards the promenade river, the pedestrian and road streets, the access from the high street, the reached from the low street, the Desert Square, the school.

Secondly, the architects have saved all the spaces that are just under the "Pormetxeta Ramp" in order to get 2,341m² of commercial and service public facility.

The third advantage is to make it work as a micro-public space: micro-squares, points of meeting, rest stands, outdoors stays, in other words to build a phenomenological sequence.

And the last advantage is that it is acting as a multifunctional element. It works as a street furniture, balustrade, pergola, lantern... creating a very equipped place.

The Animated Square

The Pormetxeta Square appears under the long ramps. There are some holes that qualify the space with playgrounds and make it become a unique topography. Some big artificial trees of 11.5m tall, that are made of corten steel, hang boulders on the cup of the tree. They work as parasols and under them somethings that people can feel are the laughters and the talks of children and the roller skaters. The Pormetxeta Square and the different levels routes, that make up at the same time a unique project and also an independent project, due to the contrast between, both intensifying their purposes.

Architecture vs. Urbanism

The architects' aim is to build a city to understand the public space and private space as a hybridization between the city planning and the edification model, where Architecture and the Urbanism are mixed at all times. Consequently the architects have planned an urban strategy as a business and dialogue platform between citizens, creating a new social cohesion.

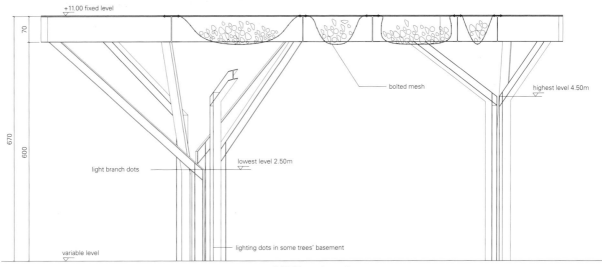

a-a' 剖面图 section a-a'

辛特拉论坛
ARX

辛特拉论坛项目位于通往辛特拉 (Abrunheira港口) 的IC-19大道附近，与Feira Nova超市相邻，是建在原有商场地块的综合性嵌入结构。

商店和影院原来位于此，之后它们就被拆除了，来为新建筑提供空间。同时超市仍然在运营，直到新建筑施工完成，在其移至一处全新的商业地块后，便转为Pingo Doce超市全新开业。

该建筑与IC-19大道之间建立的关系十分显眼，就像它与辛特拉山脉及其突出的起伏轮廓均建立起联系一样。

新建筑设计在这些环境中设立了起始点，并且旨在调节这些处于同一现代发明，却不同时空的路程。

该建筑是一个清晰的体量，沿着高速公路而建，同时环绕着一个大型地表面，被分为两个对立的垂直体量：一个是黑色混凝土制成的厚重底座，连接着建筑与地面，另一个是拥有白色饰面的体块，略向后设置，有着明显的葡萄牙传统风格。

这些材料与纹理之间的强烈对比使其整体构成形成抽象且现代的外观，并展现为优雅的文化统一体。这便是建筑师以此为出发点而设计的全新的辛特拉广场的商场空间。

传统的设计多在大小不一的广场内设置若干条笔直的室内街道，而在这个项目中，各式各样的笔直商业片区有着不同的建筑特征，它们甚至有着自己的名称 (或概念)，用来定义各处空间，以开发每处空间氛围的特殊性。

例如，使用清水混凝土建成的、拥有沉重触感的"混凝土商场"，或使用黑色玻璃、黑色石材建成的"优雅商场"，带有精致的不锈钢细部和过滤的光线。建筑师建造了一座令人称奇的"绿色商场"，是一处优雅的浮动空间，自然地承托着基础设施，利用绿色玻璃立面将人们抬升至抽象的绿色氛围中。另外还有一座"家庭式商场"，由层压木结构和涂层构成。木材拥有无可替代的触感舒适度，向人们讲述了家的感觉。

人们突然停止在此处：美食广场，在这里，花园露台将人们置于能看见辛特拉山脉的聚居地，好像这里是山脉的延伸一样。

人们还能看见其他的场景：每处场景都有自身独特的特色，将这些场景合并在一起时，能够产生明显的城市多样性，就像一座城市一样。

当这些协调一致的商场交织在一起时，其间的连续性创造了广场，作为聚会的场所。它们沐浴在自然光下，光线来自于镶嵌乳白玻璃的天窗，这些天窗形状各异，穿过天花板，盘旋在空中。在室外，这些商场的轮廓展现为独特的几何外形。固态灯在周围成为一个标志，使辛特拉

论坛成为全新的中心。大型折叠金属板(图形)准确无误地将建筑架在IC-19路大道上。这一构件体现了建筑的本质,摒弃了通常使用的不必要的装饰以及优雅的设计。因此,建筑师希望它在很长的一段时间之内仍然能够保持现代性。

Forum Sintra

The design project for Forum Sintra deals with a complex intervention on a pre-existing shopping mall adjoined to Feira Nova supermarket, near the IC-19 road arriving at Sintra(Abrunheira). The space once occupied by shops and movie theaters was demolished to give way to this new building, while the supermarket is still operating until the intervention is concluded, after which will move up to a different business ground, starting to operate then as Pingo Doce supermarket.

The relationship established with IC-19 road is quite evident, as is established with Sintra mountain range and its recognizable and waving profile.

The new architecture design had its starting points in these conditions, and it aimed at conciliating these drives, so different in time and space, in a contemporary invention.

The building emerges as a clear volume, following the freeway and curling around the large existing surface. Its mass is divided in two opposite vertical extracts: a heavy base in black concrete, connecting the building to the ground, and a second volume, pushed back, in white finish, as a clear reference to Portuguese tradition.

The strong contrast of these materials and textures at once gives the composition a radically abstract and contemporary look and installs an elegant and culturally integrated presence. This is then the architects' starting point for carving the mall spaces of a new Forum Sintra.

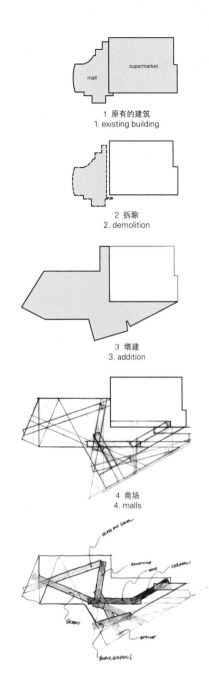

1. 原有的建筑
1. existing building

2. 拆除
2. demolition

3. 增建
3. addition

4. 商场
4. malls

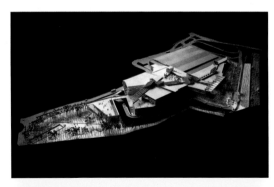

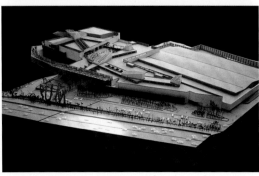

Traditionally designed as straight indoor streets coming together in multiple-sized squares, in this project the configuration started with an assortment of straight mall segments from diverse constructive identities. They even have their own work names (or concepts) used as guidelines for developing the specificity of each atmosphere.

The architects have, for instance, the heavy tectonic "concrete mall", fully made in fair-faced concrete, or the black-glass, black-stone "elegant mall", with meticulous stainless steel details and filtered luminosity. The architects have a surprising "green mall", an ethereal and floating space, bearing its infrastructures naturally and lifting people in an abstract green atmosphere, with its

东立面 east elevation

green-glass floor. They also have the "home mall" with its laminated wood structure and coating. Wood has that irreplaceable tactile dimension, speaking to people, telling people of home.
People come to a sudden halt: the food court, where a gardened terrace places people in a visual rendez-vous with Sintra mountains, as if it were its extension.
People have yet some others, each with its own individual character, and when putting together inspires a strong urban diversity, just like in a city.
When they intersect, this succession of in-line malls creates the squares as places of convergence. These are showered by natural light coming from opal glass skylights, which are oddly-shaped hovering objects piercing through the ceiling. Outside, they assume their unique unexpected geometries. These solidlamps will become a hallmark, in the surroundings, for this new centrality being created by Forum Sintra.
The presence of the large folded metal plate (graphics) unmistakably puts the building over IC-19 road. This piece embodies the architectural nature of the building, dismissing the usual superfluous decorations and straightforward elegance in its design. The architects therefore hope it stays contemporary for a long time.
ARX

北立面 north elevation

项目名称：Forum Sintra
地点：IC-19 - Rio de Mouro, Sintra, Portugal
建筑师：ChapmanTaylor España _ Arquitectura y Urbanismo, ARX PORTUGAL, Arquitectos Lda., José Mateus e Nuno Mateus
项目团队：Sofia Raposo, Ricardo Guerreiro, Emanuel Rebelo, Fábio Cortês, Diana Afonso, Miguel Torres, Filipe Cardoso, Marc Anguille, Bruno Martins, Joana Pedro, Décio Cardoso
项目经理：João Dantas
设计顾问：T&T
照明顾问：Har Hollands Lichtarchitect
景观建筑师：Jardins do Paço
工程师：Dimstrut, PEN, FICOPE / Engexpor.
承包商：Consortium MotaEngil / Opway
业主：Imoretalho - Gestão de Imóveis SA e Multi 25 - Sociedade Imobiliária
用地面积：106,750m² / 总建筑面积：47,516m² / 有效楼层面积：230,000m²
施工时间：2009—2011
摄影师：©FG+SG Architectural Photography

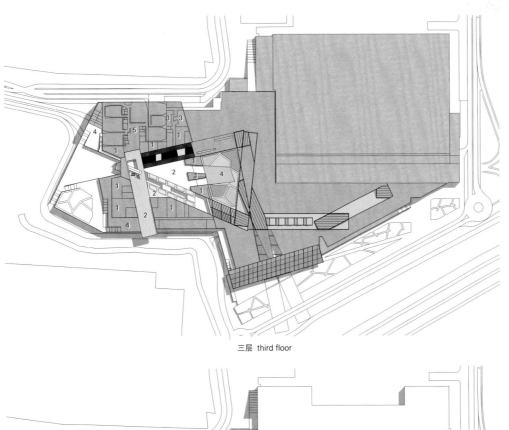

1 餐厅
2 美食广场
3 卫生间
4 室外露台
5 影院

1. restaurant
2. food court
3. toilet
4. outside terrace
5. cinema

三层 third floor

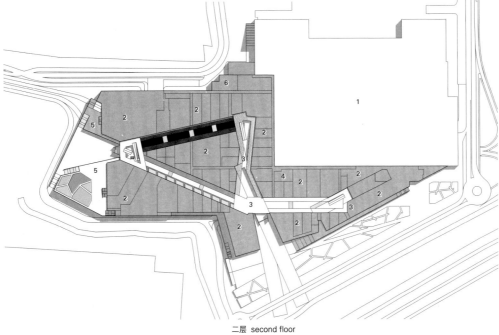

1 原有的零售区
2 新建的零售区
3 商场
4 卫生间
5 室外露台
6 管理间

1. existing retail
2. new retail
3. mall
4. toilet
5. outside terrace
6. management

二层 second floor

1 原有的零售区
2 新建的零售区
3 商场
4 卫生间
5 室外露台
6 装货间

1. existing retail
2. new retail
3. mall
4. toilet
5. outside terrace
6. loading bay

一层 first floor

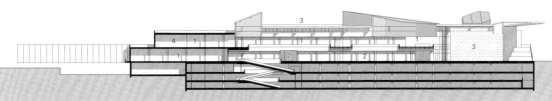

1 原有的零售区 2 商场 3 室外露台 4 管理间 1. new retail 2. mall 3. outside terrace 4. management
A-A' 剖面图 section A-A'

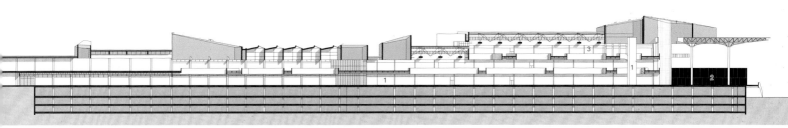

1 商场 2 室外露台 3 美食广场 1. mall 2. outside terrace 3. food court
B-B' 剖面图 section B-B'

曼福综合馆
TSM Asociados

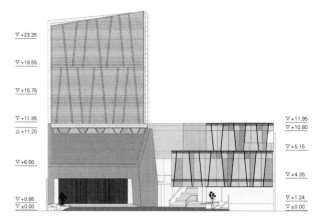

西立面 west elevation

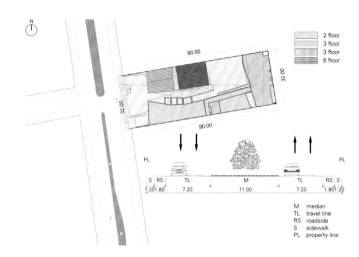

这个项目是为西班牙的曼福跨国保险公司而建造的商业综合体。其总建筑面积为7872.84m²，包含一座两层的商业办公和医疗中心大楼、一座教堂、一座六层的骨灰安置所、守灵区以及两间地下室。该建筑的主要设计原则是将所有的功能围绕着地块内的一个广场设置，以此来创造一处充满了欢迎气氛的舒适环境。

教堂位于入口的左侧，可以容纳100人。教堂作为一个中立的神殿而建。室内和室外采用的覆层的材质均为阿基雷帕生产的秘鲁石板。圣坛的背面设置了木板，木板上设有天窗，用于自然照明；而窗框上的LED灯带在没有自然照明的区域形成了永久照明的效果。

给人以沉重感的教堂建筑与骨灰安置所相得益彰，相比之下，骨灰安置所由于其带有椭球形遮阳百叶的立面而显得轻快，这样的立面使室内氛围更加私密，同时也没有对光线入口造成影响。

两层的医疗中心面向入口的右侧。它有一个连续的玻璃立面，能够将不同的玻璃类型结合起来，从浅色透明到暗色不透明。这个立面深入到综合体建筑的末端，直入办公大楼。

入口由两条人行道组成。一条作为走廊，在一层持续延伸，位于教堂和骨灰安置所的附近，直至守灵区的入口。在这片区域和地下室之间有一个垂直连接设施，即一个特殊规格的电梯，设在安置送葬队伍的区域和灵柩区域的入口。第二条人行道则是一个开放的楼梯，楼梯止于上层的广场，而广场便是曼福秘鲁办公楼入口的所在地。

Mapfre Complex

The project is a commercial complex for Mapfre, a Spanish transnational insurance company. With a total built up area of 7,872.84m², it is composed of a two story commercial office and medical center, a chapel, a six story columbarium building, a wake area and two basements. The main design principle was to arrange all functions around a square inside the property, thus creating an inviting and comfortable environment.

The chapel is located to the left side of the entrance and sits 100 people. It was intended as a neutral temple. The cladding, both internal and external, is made of Peruvian flagstone from Arequipa. The altar has a back of wooden veneer with a top skylight for

natural lighting; however, LED strings on the window frames create an effect of permanent daylight whenever there is no chance of natural lighting.

The heavy chapel volume tops off with the columbarium building, in contrast to it, the building seems light due to its facade of ellipsoid sun louvers to keep a more private atmosphere inside without compromising the entrance of natural light.

Towards the right side of the entrance is the two story Medical Center building. It has a continuous glass facade that combines different glass types going from light and transparent to dark and almost opaque; this facade enters the complex ending onto the office building.

The entrance is composed by two pedestrian accesses. One is a corridor that continues on the first level, next to the chapel and columbarium until reaching the entrance to the wake area. Between this area and the basement there is a vertical connection, a specially sized elevator for the entrance of the funeral procession area and the coffin. The second path is an open staircase that ends in the upper plaza where the entrance to the Mapfre Peru office building is located.

项目名称：Mapfre Complex
地点：Independencia, Lima 28, Perú
建筑师：Alvaro Grimaldo Tremolada, Roberto Borda Sboto
合作者：Cristina Pardo, Alessandra Peña, Jimena Serra
用地面积：2,880m² / 总建筑面积：2,056.29m² / 有效楼层面积：7,872.84m²
设计时间：2010—2011 / 施工时间：2012—2013
摄影师：©Gonzalo Cáceres (courtesy of the architect)

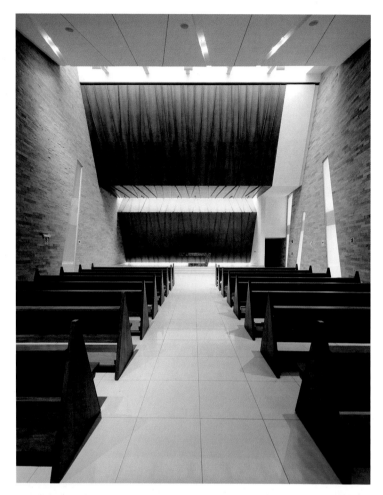

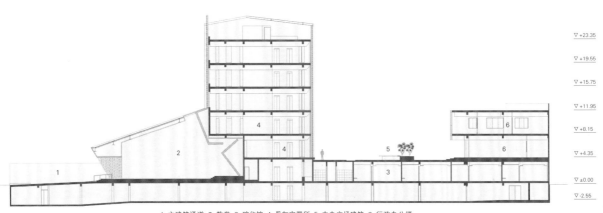

1 主建筑通道 2 教堂 3 殡仪馆 4 骨灰安置所 5 中央广场建筑 6 行政办公楼
1. main building access 2. chapel 3. funeral parlour 4. columbarium 5. central plaza building 6. administrative offices
A-A' 剖面图 section A-A'

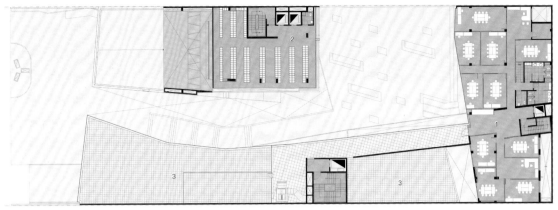

1 行政办公楼 2 骨灰安置所 3 建筑屋顶
1. administrative offices 2. columbarium 3. building roof
三层 third floor

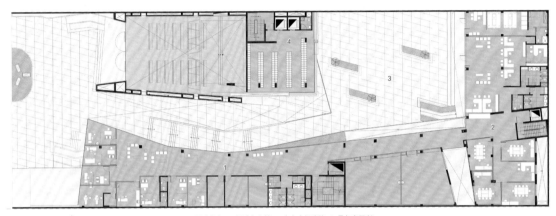

1 医疗中心 2 行政办公楼 3 中央广场建筑 4 骨灰安置所
1. medical center 2. administrative offices 3. central plaza building 4. columbarium
二层 second floor

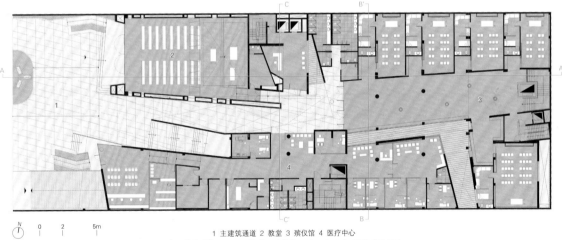

1 主建筑通道 2 教堂 3 殡仪馆 4 医疗中心
1. main building access 2. chapel 3. funeral parlour 4. medical center
一层 first floor

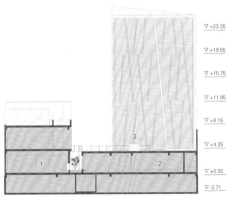

1 医疗中心 2 殡仪馆 3 中央广场建筑
1. medical center 2. funeral parlour 3. central plaza building
B-B' 剖面图 section B-B'

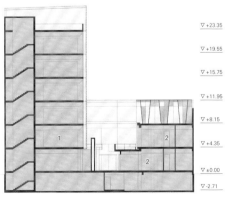

1 骨灰安置所 2 医疗中心
1. columbarium 2. medical center
C-C' 剖面图 section C-C'

日式城市住宅
Japanese Urban Dwell

Traditional Modernity

传统与现代

  日式风格是当下唯——种仍然保持特定民族精神和特性的建筑风格。在当代建筑作品的不同风格和语言当中，我们总能轻松地解读出日本建筑风格的独创性。结合了独创性和诗意般轻盈性的简约主义是日本风格的主要特征。这种大胆的建筑风格似乎找寻到成就其标志性和忘我风格的途径。而这种理解方式当中的某些东西，试图影响和产生与当下时代精神非常契合的建筑风格。日式建筑风格除了在公众当中获得普遍赞赏之外，日本建筑师也在过去两年中连续获得了普利策建筑奖的殊荣。

  除了大师级建筑师的作品外，小型以及一些名不见经传的建筑师事务所中也产生了很多建筑设计作品。这些建筑作品的规模都属于分离的独立家庭住宅。

  因建筑风格是设计师的态度和委托人的接受度共同作用的结果，因此住宅建筑可以说是代表一个场所精神的最纯粹的方式。然而，日本建筑师采取的大多数设计解决方案可能不会被世界其他地区的建筑师所接受。

Japanese is nowadays the only kind of architecture which can still maintain a specific national defined spirit and character. Amongst different styles and languages of contemporary production we can always and easily read the originality of Japanese architecture. Minimalism is the main feature associated to Japanese style, but with a particular twist of originality and poetic lightness. The daring architecture seems to have found a way to be iconic and yet selfless. There is something within this way of understanding, approaching and producing architecture which particularly matches the zeitgeist of our times. Beside the general appreciation Japanese architecture earns a reputation amongst the public we had also the prestigious award of the Pritzker Prize assigned for the consecutive past two years to a Japanese architect.

Beside the work of the Masters, there is a rich production from small scale and not yet famous architectural offices. The scale of this production is that of detached single family house.

Because architecture is a synergy of designer's attitude and client's acceptance, the residential architecture can be considered as the purest for representing the spirit of a place. Most of design solutions adopted by Japanese architects would probably not be accepted in other parts of the world.

K住宅_House K/Hiroyuki Shinozaki Architects
长廊住宅_Promenade House/FORM/Kouichi Kimura Architects
白砂住宅_Shirasu House/ARAY Architecture
Okusawa住宅_Okusawa House/Hiroyuki Ito Architects+O.F.D.A.
山外小屋_Beyond the Hill/acaa/Kazuhiko Kishimoto
雪松住宅_House of Cedar/Suga Atelier
螺旋之家_Life in Spiral/Hideaki Takayanagi Architects
阳台之家_Balcony House/Ryo Matsui Architects

日式城市住宅——传统与现代_Japanese Urban Dwell – Traditional Modernity/Michele Stramezzi

日本建筑

日本建筑首要并且最显著的特征在于它是当代建筑中唯一一个非本土的建筑类型，但这种建筑类型却仍能被认可和明确界定。例如，当我们无法辨别一项法国建筑设计是出自匈牙利还是美国的建筑师时，以及当非洲或亚洲半传统建筑作品无法定义单个民族风格时，我们都会注意到日本的大部分建筑作品如何轻易地被归属为当代建筑意图和设计的一种特殊方式。

日本当代建筑常被认为规模极小并且设计风格简单，这与其他建筑类型，例如斯堪的纳维亚建筑风格的特征相同；然而日本建筑的简约方法常常饱含深意并充满诗意般的温暖。符号的缺失并没有减少其空间的意义，相反却常常能在丰富性、生活和整体复杂性方面形成非凡的深度。带有整合语言和材质面板的日本建筑作品非常时尚并且具有当代特性。即便或多或少地应用了传统技术和材料，设计的结果也会完全与最前沿的当代建筑风格相一致。

日本人的人生观崇尚去除和削减掉无关紧要的事物后显现出的一种增强的单纯美感。——安藤忠雄

简约但是创新和大胆的日本建筑似乎已经找到成就其标志性和忘我风格的方法，这是一个再次强调其独特性的特征，并且将被证实成功地表达了2009年经济危机之后世界的时代精神。当代日本建筑的第二个特性其实是从批评家和公众中获得的国际方面的成功。例如，在过去的两年中，两名日本建筑师获得了普利策建筑奖：2014年的坂茂和2013年的伊东丰雄。而从2010年至今，我们还需要将妹岛和世和西泽卫立双人设计组加入到获奖名单当中：五年内六位获奖者，同时，妹岛和世也是除了扎哈·哈迪德之外第二位获此殊荣的女性，并且她是2010年唯一一位经选举担任威尼斯国际建筑双年展主管的女性。美国和日本各自拥有七位普利策获奖者，向我们展示了其是如何引领获奖国家名单的。值得一提的是，该奖项曾经连续三年由美国建筑师获得，而这在某种程度上也反映了当代日本建筑的成功。

Japanese

The first and most remarkable character about Japanese architecture is that it is the only non-vernacular type of contemporary architecture which can still be recognized and clearly identified. When we can not distinguish the building design of a French, for example, from a Hungarian or American architect, when even the production of African or Asian semi-traditional architecture wouldn't define a single national style, we all notice how the most of the Japanese production can be easily ascribed as belonging to a peculiar way of intending and designing contemporary architecture.

Japanese contemporary architecture is often perceived as minimal and simple, a feature common to other kinds of architecture style, the Scandinavian for example; and yet the minimalistic approach of the Japanese is always full of meaning and poetic warmth. The absence of signifiers doesn't reduce the meaning of this space but on the contrary always results with its extraordinary depth in richness, life and overall complexity. The Japanese production is extremely up to date and contemporary with its composition languages and material palette. Even where more or less traditional techniques and materials are applied, the result is absolutely coherent with the most up to date contemporary architecture.

The Japanese view of life embraced a simple aesthetic that grew stronger as inessentials were eliminated and trimmed away. – Tadao Ando
Minimalistic and yet innovative and daring Japanese architecture seems to have found a way to be iconic but selfless, a characteristic which again underlines its uniqueness and will prove extremely successful in expressing the zeitgeist of the world after the crisis of 2009. The second peculiarity about contemporary Japanese architecture is in fact the international success which is attaining both from critics and public. For the past consecutive two years, for example, the Pritzker Prize awarded two Japanese architects: Shigeru Ban in 2014 and Toyo Ito in 2013. From 2010 till today, we need to add to the list the duo Kazuyo Sejima & Ryue Nishizawa: six laureates over five years. Kazuyo Sejima is also the second woman ever, beside Zaha Hadid, who has been awarded with the Prize and, in 2010, the one and only woman ever selected for the position of Director of the International Venice Biennale Exhibition of Architecture. The Pritzker Laureate list shows us how both the United States and Japan lead the list of winning coun-

依时间顺序研究这两种平行的状况是一件有趣的事情。凯文·洛奇、贝聿铭和理查德·迈耶分别于1982年、1983年和1984年获得奖项。由这些设计师设计的玻璃和钢结构建筑映射了柏林墙倒塌:那是一个经济突飞猛进,并且西方发达国家、北美洲和欧洲国家以及日本繁荣发展的年代。

日本建筑+栖居

20世纪90年代象征性建筑当属毕尔巴鄂的古根海姆博物馆的极具标志性的大楼:事实上,体现经济成就最明确的形式是博物馆,它甚至比高大的玻璃和钢结构办公大楼更能体现经济成就的荣耀。21世纪第一个十年期间的标志性建筑是位于北京的CCTV总部,它是全球化世界的象征,由首位推断这种现象的建筑师设计而成。2009年世界经济危机过后,建筑界正寻找一种新的节制方式并形成更加合理的规模:过去时代遗留下来的大型项目和过于标志性的建筑显得有些怪异和落伍。

2000年日本经历了一次经济衰退,它比世界其他国家经济衰退的起始时间早了十年之久。这次经济萎缩并没有形成穷困的状况,而事实上,据报道它反而增加了人们的幸福感。日本建筑作品似乎崇尚时代对质朴和意义、规模感、对理解空间和材料的灵性的一般需求以及在不丢失传统和特性的情况下崇尚不断超越极限的能力。除了名家的设计作品之外,日本还有众多年轻且富有创新能力的建筑设计师,他们在不断地重组民族的建筑语言。鉴于明确的经济原因,这些更容易发生在独户住宅的设计中。

建筑是设计师和委托人共同作用的结果:而最能表达一个场所精神的建筑当属住宅建筑。然而日本建筑师大量采取的特定设计方案如果没被彻底否决的话,也很难被世界其他区域轻易接受。

日本建筑+栖居+城市

日本地形大多并不平坦,且山脉崎岖或高低不一。因领土的地震性

tries with seven laureates each. There has been a time when the Prize went for three consecutive years to an American architect, mirroring somehow the contemporary Japanese success.

It is interesting according to time to contextualize these two parallel situations. Kevin Roche, Leoh Ming Pei and Richard Meier won their Prizes respectively in 1982, 1983 and 1984. Their architecture of glass and steel matches the glory of a period which will lead to the fall of the Berlin Wall: years of economical growth and prosperity for the countries of the western developed world, North America and Europe with Japan.

Japanese + Dwell

The symbolic architecture of the nineteen-nineties is the hyper-iconic building of the Guggenheim Bilbao: the definitive glorification of economical success is in fact a museum, even more than a tall office tower of glass and steel. The icon of the first decade of 21st century is the CCTV Headquarter in Beijing, a symbol of the globalized world, designed by the first architect who ever theorized that phenomenon. After the 2009 world crisis, the world of architecture is looking for a new sobriety to form more reasonable scales: mega projects over iconic architecture appear like bizarre and anachronistic leftover of a past era.

Japan in 2000 is going through an economical recession which started already ten years earlier than the rest of the world. The result of the economy contraction is not leading to misery and actually is reported to increase the perception of happiness amongst people. The Japanese architectural production seems to embrace the general need of modesty and meaning, the sense of scale and spirituality in the understanding of space and material to period, and the ability to keep pushing the limits without losing the contact with tradition and identity. Beside the work of the famous masters, there is a variety of young and creative small architecture designers which are constantly reinventing the language of the national architecture. The scale at which all this can happen, for clear economical reason, is that of the single family house.

Architecture is a synergy of designer and client: within residential architecture the expression is the purest spirit of a place. Specific design solution so freely adopted by Japanese architects would not easily be accepted in other parts of the world, if not completely refused.

长廊住宅，滋贺县
Promenade House, Shiga

质，高度的增加并不是传统的建筑方式，也不是最为经济的设计方案。

拥挤的空间常常形成狭小的地块尺寸。小型住宅建筑的地块经常形成狭长的不规则形状。建筑物的规模时常受到场地形状的影响，高度也经常受到临近建筑的约束。我们将介绍几处在极小和狭窄地块中设计的具有独特类型的房屋实例。

此种空间上的不足形成如此靠近彼此的独立式住宅。城市建筑经常是内向型并且对外封闭。房屋开口一侧的设计可以自由选择——采取视觉上的全封闭或是全透明设计。

由松井亮建筑都市设计事务所设计的位于东京的阳台之家是一栋带有四层空间的独立住房。地块呈常规的长矩形，而该独立建筑四面中的三面都与另一栋建筑保持最小距离。该房屋在这三面形成封闭状态，而在面向街道的短边上完全开放。2m深的阳台起到了缓冲住宅与城市之间的关系的作用。该入口元素的超大进深，连同阳台上特别种植的小树，似乎在深度上增加了进出位置的层次感，形成家庭生活中隐私性和公开性的变化。住宅采用的材质面板都是基本和简单的材料，包括轻质混凝土、建筑外围护结构的玻璃和地面、阳台和屋顶使用的木料。

由acaa建筑研究所的建筑师岸本和彦设计的位于横滨的山外小屋是一栋结合公共与半公共小画廊和办公空间的私用住宅建筑。建筑的第一层采用开放型设计，并且允许行人穿过，不同楼层的室内庭院与地块的斜坡性质相呼应，并且调节公共和私人区域的不同梯度；两个独立楼梯通往办公室以及其他两层的生活区。楼体仿佛一个屹立于延伸的底层架空柱之上的封闭木质盒式结构，木材将盒体整体包围，建筑师没有设计任何面向街道的洞口，建筑简单的体量在某个角度被切掉，并且其上镶嵌玻璃，来为建筑提供光线。

因此，材质面板由楼体使用的木材和庭院的景观、底层架空柱的轻质混凝土、室内面向内院立面的玻璃和全白灰泥组成。作为表面的底层架空柱似乎在指引街道、室内庭院和毗邻林木区域之间的风景和空间。

Japanese + Dwell + Urbanity

Japanese topography is mainly not flat but rather mountainous of varying heights. Because of the seismic nature of the territory, to grow in height is not the traditional way of building and not the most economical.

The congested space leads to dimension of the lots which are usually very small. Often the plots for small residential architecture are of irregular shape, long and narrow. The footprint of the building is often dictated by the shape of the plot. The height is also usually regulated to be conformed to neighboring buildings. There are several famous examples of overly small and narrow plots which are still filled in with unique type of houses.

The result of this lack of space brings residential buildings which are free standing so close to each other. The urban behavior of the architecture is often extremely introverted and closed towards the outside. The open side of the house leaves the freedom of the choice, fully closed or completely transparent to the sight.

Balcony House in Tokyo, by Ryo Matsui Architects, is a single family house which develops on 4 multiple levels. The size of the plot is the usual long rectangular, and the building is a free standing with a minimal distance to next buildings on three of the four sides. The house reacts being closed on this three sides and fully open on the fourth short side, with one facing the street. Two meters deep balconies, buffer the relationship that the house maintains with the city. The extraordinary depth of this threshold elements is combined with proper small trees planted in the balcony itself, seemingly increasing the layers in depth of the in and out situations, and the gradients of private and unveiled within the family life. Material palette is basic and simple and includes light concrete, glass for the envelope and wood for the floors, balconies and roof top.

Beyond the Hill in Yokohama, by acaa/Kazuhiko Kishimoto is a residential building which combines the private function of dwelling with the public and semi-public of a small gallery and office space. The first level is open and permeable for a crossing of the pedestrian public, and an inner courtyard on different levels reacts to the slope nature of the plot and modulates different gradients of public and private; two independent staircases lead to the office and to the living areas at the next two floors. The body of the house is a closed wooden box standing on stretched pilotis,

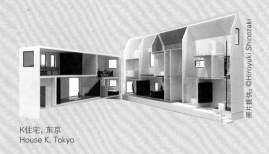

K住宅，东京
House K, Tokyo

由FORM/木村浩一建筑研究所设计的位于滋贺县的长廊住宅必须融入到一处延伸的且非常狭长的场地当中。建筑的体量是一栋嵌有两个完全封闭塔楼的简单的盒状建筑。光线由这两个塔楼以及前部和后部的洞口进入建筑。但该建筑与其自身毫无趣味性的环境并不是没有任何的视觉关系。一扇透明的后门面向一片绿地。材质面板由灰色的百叶窗，即前面景观的轻质混凝土，暗色和白色灰泥（短的一侧使用暗色灰泥，而长的一侧则使用白色灰泥）组成。

由ARAY建筑事务所设计的位于鹿儿岛的白砂住宅拥有一个非典型的轮廓，它不是常规的长矩形，而是一个带有尖角的地块，形成一个带有切边的方形轮廓。而最终的设计结果是一栋外观奇特并且呈现带有倾斜表面的城市塔状建筑外形的独院住宅。材质面板以统一的灰色表面材料来平衡建筑体量富有表现力的处理。该建筑另一个特有的元素是分为3个梯度的立面洞口系统：全开放、部分开放和封闭，当然，半堵砌砖墙允许视线和风的进入，在房屋和其内外部的生活之间形成一种中间关系。最后一个令人称奇的元素是客厅空间内面向外部的大型洞口——它在室内和户外空间（一个由两条小城市车道交叉而界定的小型前花园）之间建立一种完整的互动关系。

由筱崎弘之建筑设计事务所为两户人家设计的位于东京的K住宅是一栋复式住宅，建筑师将它描述为是一处恬静的居住区。建筑的轮廓呈典型的延长和狭窄状；该建筑分为三个相互连接的不同细长体量，它们不呈一条直线并且高度各异。三个体量中的一个带有陡峭的斜坡屋顶。立面窗户元素呈不对称状并且故意采用参差不齐的设计。材质面板通过在两侧立面和屋顶使用全白色灰泥的统一处理来平衡体量具有表现力的处理。

由高柳英明建筑研究所设计的位于东京的螺旋之家有一个不规则的四边形轮廓，其中两侧紧邻附近的建筑，而另外两侧则呈开放状。建筑体量的顶部呈折叠状，并带有斜线切面。该建筑的材质面板是基础的轻质混凝土，嵌入地面，而建筑主体立面交替使用白色灰泥和玻璃材

and the wood fully wraps the box, with no openings towards the street. The simple volume is chopped off on an angle and glazed so to provide the light into the building.

The material palette therefore consists of wood of the body and the landscape of the courtyard, light concrete for the pilotis, glass and full white plaster for the internal elevations, those facing the inner court. The pilotis are surfaces, seemingly directing the flow of people views and space between the street, the inner courtyard and the adjacent trees area.

Promenade House in Shiga, by FORM/Kouichi Kimura Architects, has to fit into a stretched, long and very narrow site. The volume is that of a simple box with the insertion of two towers, totally enclosed. The light comes in the building from the two towers and from the front and back openings. The building doesn't practically have any visual relationship with its own rather uninteresting surroundings. The back door has a transparency towards a green spot. The material palette consists of shades of gray: that of light concrete of the front landscape, the dark and white plaster, dark on short side, white on the long.

Shirasu House in Kagoshima, by ARAY Architecture has an atypical footprint, not the usual long rectangular but a squared corner plot, and leads to a squared footprint with a cut edge. The result is an odd looking single family house with the shape of an urban Ziggurat with inclined facade surfaces. The material palette counters balance the expressive treatment of the volume with a tone of gray facade material. Next peculiar element of this building is the facade opening system in a gradient of three: full and semi open and closed, of course, half full masonry walls permeable to sight and wind, give a condition of intermediate relationship between the house and the life running inside and outside. Last and somehow surprising element is the big opening of the living room space towards the outside, which establishes a complete interrelation between the indoor and outside space, a small front garden defined by the crossing of two small urban driveways.

House K for Two Families in Tokyo, by Hiroyuki Shinozaki Architects, is a duplex house in what it is described to be a quiet residential area. The footprint of the building is typical stretched and narrow; the volume is articulated into three different elongated volumes not aligned and varying in height. One of the three volumes finishes with a steep pitched roof. The composition of the

Okusawa住宅，东京
Okusawa House, Tokyo

料。通过开放两侧的两个全高玻璃立面，建筑达到形成完整视觉互动的可能性。条纹窗帘系统调节隐私度，在城市与各自日常生活之间形成每个家庭都渴望达到的干预效果。

由伊藤裕之建筑师事务所和O.F.D.A.设计的位于东京的Okusawa住宅同样带有一个不规则的四边形轮廓以及一个相当连贯的建筑体量：立面和体量需求都适用临近建筑，以维持其原有的日光、通风和隐私状况。建筑一层的立面对街道完全封闭，临长街的立面上的窗户采用极为不规则的式样。建筑三层和四层的较短一侧带有洞口。材质面板以完全裸露的混凝土的处理方式再一次平衡建筑体量具有表现力的处理，使建筑的立面和屋顶保持一致。

由须贺工作室设计的雪松住宅呈隧道状，由地面突起的二层小楼发展而成。隧道较长的一侧完全封闭，较短的一侧对外开放并且全部镶嵌玻璃，其中一侧起到面向附近河岸开放的视觉浏览器的作用。

这栋非常简单的建筑物外部没有什么特殊设计，但却内有乾坤：外围护结构不仅包围房间，也将外部覆顶和封闭空间包覆，其中包括一座花园和一个穿插于生活空间和外部合理都市生活之间的露台。

window openings on the facade is asymmetric and intentionally carefully unaligned. The material palette of the design counters balance the expressive treatment of the volume with a full uniform treatment of white plaster for both elevations and the roof. Life in Spiral in Tokyo, by Hideaki Takayanagi Architects, has an irregular quadrilateral footprint very close to the adjacent buildings on two sites, free on the other two. The volume is folded on the top with diagonal cuts. The material palette of the building is basic light concrete for the insertion to the ground, and alternates white plaster and glass for the body facades. The building choses for the possibility of a full visual interaction with its context through its two full glass facades on free sides. A system of stripe curtains modulates the gradient of privacy which the family will desire to interpose between the city and their own daily life.

Okusawa House in Tokyo, by Hiroyuki Ito Architects+O.F.D.A. has also an irregular quadrilateral footprint and quite an articulate volume: elevation and masses need adjust to the adjacent residential building to preserve its original conditions of sunlight, ventilation and privacy. The ground floor facade is completely closed towards the street, and the window pattern rigorously irregular on long street facing facade. The short sides have openings on second floor and third floor. Material palette counters balance once again the expressive treatment of the volume with a fully exposed concrete treatment, which unifies the building volume's elevations and roof.

House of Cedar in Osaka, by Suga Atelier, has a tunnel shape, developed on two floors, slightly raised from the ground. The long sides of the tunnel are completely closed, and the short ones are open and fully glazed, one of them performing as a visual cannon towards the nearby river banks.

What is not happening on the outside of this extremely simple construction is happening inside: the envelope is not wrapping just rooms but an extra outdoor covered and enclosed space, and a garden and a terrace are included and interposed between the living spaces and the outer proper urban life. *Michele Stramezzi*

K住宅

Hiroyuki Shinozaki Architects

这是一套复式住宅,位于东京郊区一处安静的住宅区内。该住宅的主题是怎样将一座房子用于两个家庭使用。

位于细长的场地内的K住宅被分为三个部分:木体量、混凝土体量以及连接前两个部分的走廊。

这是因为在日本,复式住宅通常被划分为两处空间,用于保证每个家庭的私密性。但是从未来长远的角度来看,这并不是一个合理的设计理念。

虽然住宅不能轻易地被改造,但是居住于此的人们将不可避免地更换。所以这座住宅并没有分割成两个家庭的生活空间,而是设置了起居空间和辅助空间。这两种不同的空间由一条走廊进行连接,而走廊给人们留下了室外空间的印象。

木体量设有辅助空间:厨房、浴室、库房、壁橱。它宽1.8m,长16.5m,高9m,由小型木横梁和胶合木建成。出于造价和可改建的特性,建筑师建造了这个木体量,它看起来就像一个小棚屋一样。

混凝土体量则设有居住空间,包括起居室、餐厅和卧室。起居空间坚持为人们的舒适性而建。这也是一个细长的体量,宽2.7m,长13m,高5.6m。

在结构方面,沉重的混凝土体块支撑着轻质的木体量,来抵消水平力,抵抗地震和风的力量。这两个细长体量的中间为走廊,使简单的住宅内拥有了像迷宫一样的长路线。因此出于隐私性的考虑,人们能够轻易地保持一定的距离。

因此,这座复式住宅的理念,与只是简单地将一座住宅分成两座小住宅的老旧理念完全不同。

House K

This is a duplex house in a quiet residential area, suburb of Tokyo. The theme of this house is how to divide a house for two families. This house is composed of three parts, on the slender site: wooden box, concrete box and the corridor which is connecting those two boxes.

This is because usually in Japan, duplex house is divided into mainly two spaces for each of families' privacy issue. But it is not the rational idea from viewpoint of the long future.

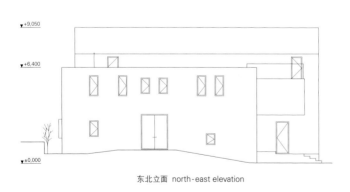

东北立面 north-east elevation

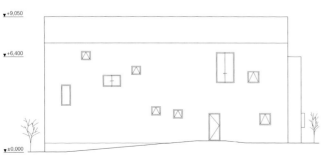

西南立面 south-west elevation

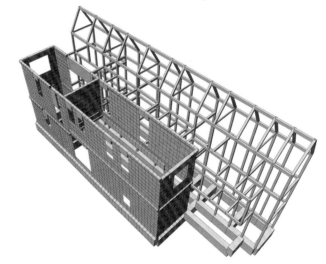

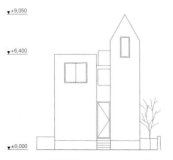

西北立面 north-west elevation

Although the house cannot be changed easily, people living in the house will change inevitably. So this house is not divided into two family spaces but is consisted of living space and supporting space. And those two different spaces are connected by the corridor which gives an impression of outdoor space.

The wooden box has supporting space: kitchen, bathroom, stock, closet. It is 1.8m wide, 16.5m long and 9m high. It was built with small wooden beam and plywood. The wooden box is built simply for cost and for the modifiability and it's like the shed.

And the concrete box has living space such as living room, dining room and bedroom. The living space is made firmly for people to stay comfortable. It is also a long and thin box, 2.7m wide, 13m long and 5.6m high.

In respect of the structure, the heavy concrete box supports the light wooden box against horizontal force, earthquake and wind. The composition of two slender boxes sandwiching the corridor makes long distance like a labyrinth in the simple house. Therefore people are easily able to make distance for their privacy.

So this is a new idea of duplex house, not old one which is simply dividing one house into two small houses.

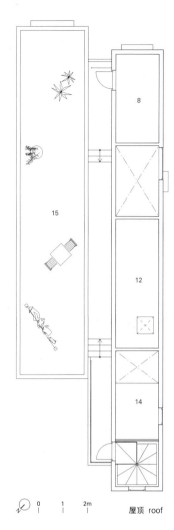

屋顶 roof

1 入口	6 卧室	11 媒体间
2 卫浴间	7 走廊	12 壁橱
3 卫生间	8 存储室	13 日光房
4 厨房	9 日式房间	14 阁楼
5 餐厅	10 起居室	15 屋顶露台

1. entrance 6. bedroom 11. media room
2. sanitary room 7. corridor 12. closet
3. W.C. 8. storage 13. sun room
4. kitchen 9. tatami 14. loft
5. dining 10. living 15. roof terrace

一层 first floor

二层 second floor

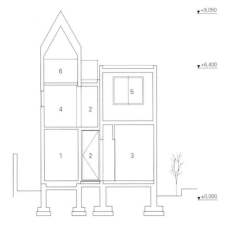

1 浴室 2 走廊 3 存储室 4 媒体间 5 卧室 6 阁楼
1. bath 2. corridor 3. storage 4. media room 5. bedroom 6. loft
A-A' 剖面图 section A-A'

1 入口 2 浴室 3 卫浴间 4 卫生间 5 厨房 6 餐厅 7 卧室 8 日光房
9 壁橱 10 媒体间 11 阁楼 12 存储室
1. entrance 2. bath 3. sanitary room 4. W.C. 5. kitchen 6. dining
7. bedroom 8. sun room 9. closet 10. media room 11. loft 12. storage
C-C' 剖面图 section C-C'

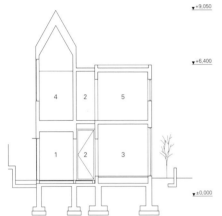

1 餐厅 2 走廊 3 起居室 4 日光房 5 卧室
1. dining 2. corridor 3. living
4. sun room 5. bedroom
B-B' 剖面图 section B-B'

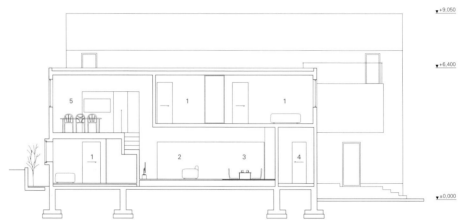

1 卧室 2 起居室 3 日式房间 4 存储室 5 餐厅
1. bedroom 2. living 3. tatami 4. storage 5. dining
D-D' 剖面图 section D-D'

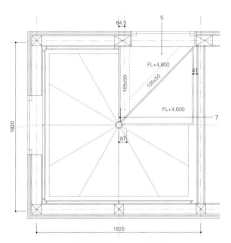

阶梯23~24　step no.23~24

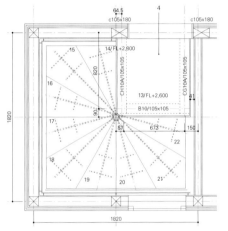

阶梯13~22　step no.13~22

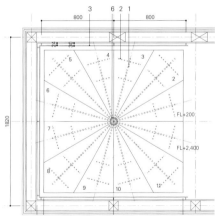

阶梯1~12　step no.1~12

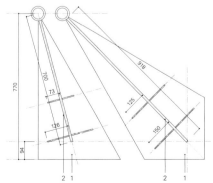

阶梯构件详图　step unit detail

1. St FB - 12 × 75
2. St Rib. PL - 6 × 50 (ss400)
3. step /
   rubber laminated timber
   St PL - t6 (ss400) oil paint
4. handrail / FB - 6 × 20 oil paint L = 1749 × 3
   bracket / FB - 6 × 20 W24 @800
5. landing step /
   structural plywood t = 24
   wood flooring t = 12
6. St 76.3ø × 9.0 oil paint
7. St PL - 4.5 × 50 L150

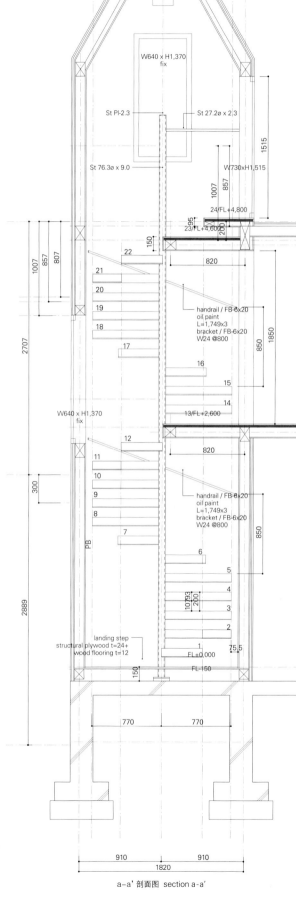

a–a' 剖面图　section a-a'

项目名称：House K
地点：Tokyo, Japan
建筑师：Hiroyuki Shinozaki
结构工程师：Tatsumi Terado Structural Studio
承包商：Sinei, Ltd
用地面积：105.13m² / 总建筑面积：161.47m²
结构：reinforced concrete, wood frame
竣工时间：2011.11
摄影师：©Kai Nakamura (courtesy of the architect)

长廊住宅
FORM/Kouichi Kimura Architects

该住宅是为一对年轻的夫妇而建造的，建于长35m、宽4m的独特地块上。

场地的外形限制充分反映在住宅的室内构成中。

该建筑长27m，宽2.7m，与狭窄的场地相匹配，以形成其轮廓。

室内空间规划了一条长长的且狭窄的走廊，人体利用走廊就能感知场地的几何轮廓。而当人们在走廊内继续前行时，便会看见几处空间蔓延开来。

长长的走廊从一层的入口便开始延伸，由脚灯引领，穿过餐厅和起居室，最后在末端连接抬升的书房。穿过书房的大型窗户，人们能够看见田园般的景色。而从入口处开始的锥形视野在此全面敞开。

在二层，两条走廊计划从有脚灯的楼梯处开始延伸，一条走廊的墙体为绿色，旨在带来色彩效果。鲜明的绿色走廊围绕着阳台，赋予附近的浴室和卫生间一种干净的印象。另外一条走廊从儿童房开始，穿过卧室，直至露天空间内的小桥处。这条走廊的设计目的是用来控制光线的，而穿过（将儿童房隔开的）透光窗帘的光线，亦或是露天空间的高侧窗射来的阳光，将引领人们前行。走廊的末端变成一座桥，安装于此的梯子连接着上下层空间，成为一个连接纽带。

绿色墙体设置在建筑的两个末端，使其整体的长度给人的印象更加深刻。这座住宅内设置的走廊都是赋予场地的几何外形强烈印记的步道。

Promenade House

The project is for the house owned by a young couple and is planned at the unique site of 4 meters wide and 35 meters long. The geometrical restriction of the site is reflected in the internal composition of the house.

The building, with a width of 2.7 meters and a total length of 27 meters, is laid out in accordance with the narrow site to draw its outline.

The internal space has been planned to have a long narrow hallway, with which human body senses the site geometry. As people proceed along the hallway they will see the spaces spreading out one after another.

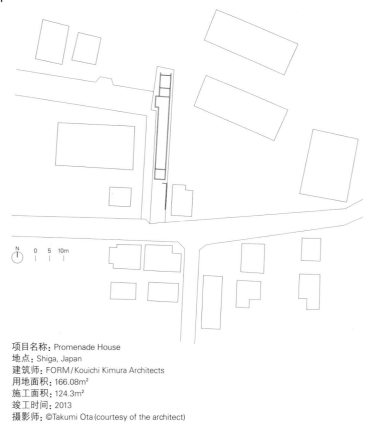

项目名称：Promenade House
地点：Shiga, Japan
建筑师：FORM / Kouichi Kimura Architects
用地面积：166.08m²
施工面积：124.3m²
竣工时间：2013
摄影师：©Takumi Ota (courtesy of the architect)

南立面 south elevation 东立面 east elevation

北立面 north elevation 西立面 west elevation

0 2 5m

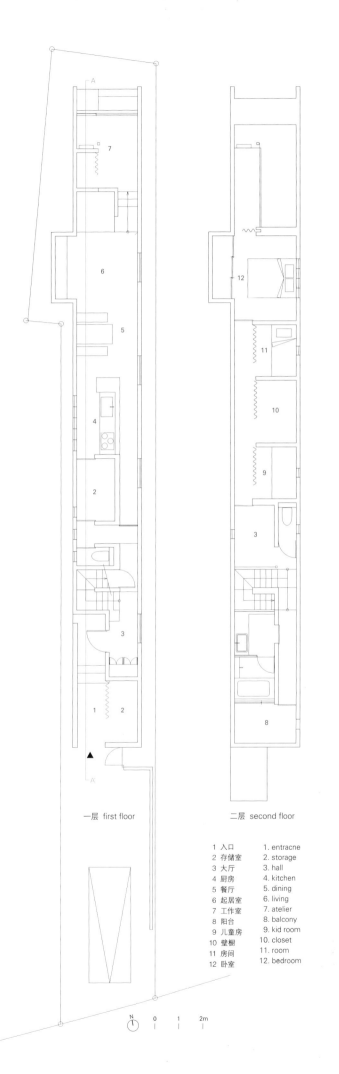

一层 first floor 二层 second floor

1 入口	1. entrance
2 存储室	2. storage
3 大厅	3. hall
4 厨房	4. kitchen
5 餐厅	5. dining
6 起居室	6. living
7 工作室	7. atelier
8 阳台	8. balcony
9 儿童房	9. kid room
10 壁橱	10. closet
11 房间	11. room
12 卧室	12. bedroom

The long hallway is extended from the entrance on the first floor, led by the footlight through the dining and living rooms, and connected to the raised study at the very end. It reaches to the idyllic view seen through the large opening of the study where the tapered line of sight from the entrance is opened up.

On the second floor, two hallways are planned to extend from the staircase that has a top light. One has a green wall aiming for color effect. The vivid green hallway surrounds the balcony, giving an impression of cleanliness to the adjacent bathroom and washroom. The other is connected from the kid room through the bedroom to the bridge at the open-ceiling space. It was designed to control light; the light through the light transmissive curtain separating the kid room, or the sunlight from the high-side light in the open ceiling space leads people forward. The end of the hallway becomes a bridge, and the ladder installed there connects the upper and lower spaces to produce continuity.

The green wall is used at both ends of the building, providing more impressiveness of the total length. The hallways laid out in this house are the promenades that strongly impress the site geometry.

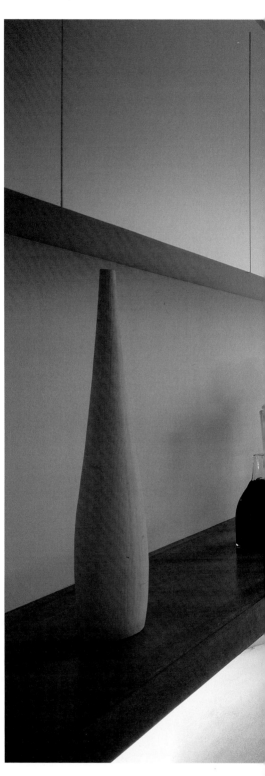

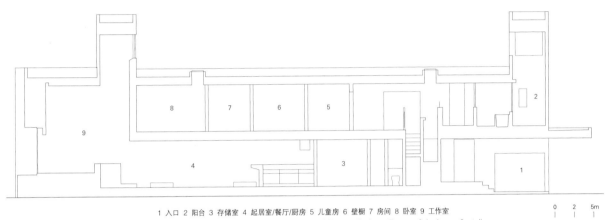

1 入口 2 阳台 3 存储室 4 起居室/餐厅/厨房 5 儿童房 6 壁橱 7 房间 8 卧室 9 工作室
1. entracne 2. balcony 3. storage 4. living/dining/kitchen 5. kid room 6. closet 7. room 8. bedroom 9. atelier
section A-A'

白砂住宅
ARAY Architecture

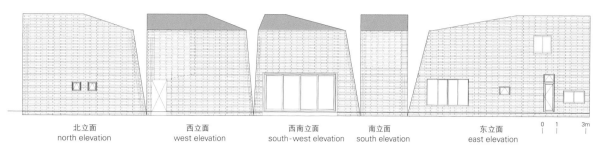

北立面	西立面	西南立面	南立面	东立面
north elevation	west elevation	south-west elevation	south elevation	east elevation

火山泥砌砖建成的可持续性住宅

该住宅位于九州南部的一处温度高、湿度大的区域内,但是它并没有依靠能源来建造,而是希望利用环境来塑造生态型生活。人们可以在这片土地上感知风的存在,进行雨水存储,并且还可进行园艺活动。这是一种本地化的生活。场地是一处位于鹿儿岛中心的附近,且向白砂高地延伸的住宅地块。之后,建筑师便将当地的住宅(可持续住宅)和泥土(白砂)作为原材料,它们通过白砂住宅来形成这片具有当地特色的高地。

这座住宅还有许多其他的地域特点,如防火、绝热、湿度条件、蓄热和明亮性等。当对场地进行密封和施工时,建筑师充分利用了单一体块的建造技术,来对开始在城市内蔓延的道路进行铺设。而整个白砂住宅施工所运用的体块都是第一次使用。原矿石嵌入室外墙体内,以提高内部体块中水分的吸收性和解吸性。同时,白砂住宅的特点是成为表达记忆的载体。住宅成为一处用泥土围起来的空间,如同窑洞一样。

室内和室外双层中空墙体减少了室内的热负荷。此外,室内墙体砌块像住宅内房间的饰面一样,围绕起来,以调整室内湿度。因此,夏季室内凉爽,冬季温暖。这是一处全年恒温的环境。它规划了白砂高地地下积累的资源的能源性能,对火山泥砌砖进行了再利用,是能够进行再循环的新型环保可持续性住宅。

此外,白砂住宅的绿色屋顶和轻质砌块的材质都是泥土。因此,它可以减少屋顶的荷载。白砂住宅的绿色屋顶也能提高住宅能效,同时能够对城市环境质量产生积极的影响。

现在,居民正在享受带有亲切氛围的自然的生活,他们没有安装空调,并且使白砂住宅的墙体和屋顶远离热环境。

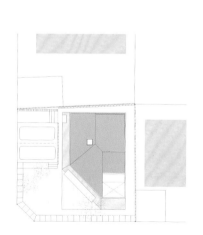

白砂住宅体块和砌砖的生产
Shirasu block and masonry production

板块切割
cutting the plate block

穿过钢筋的孔洞的加工
processing the through-hole for reinforcement bar

室内墙体体块　　室外墙体体块
inner wall block　　outer wall block

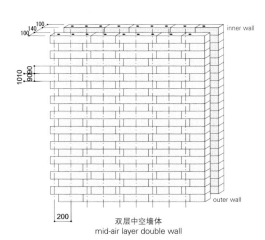

双层中空墙体
mid-air layer double wall

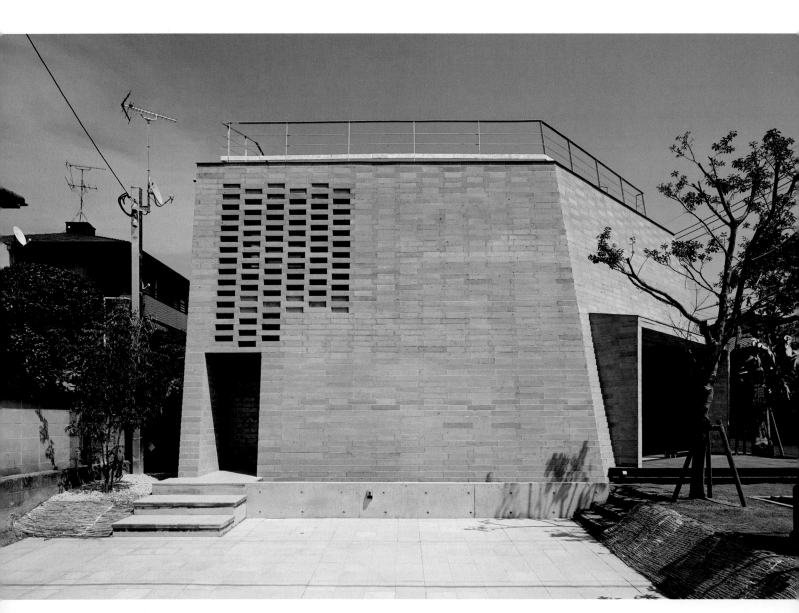

Shirasu House with Volcanic Soil Brickwork

The resident did not rely on energy in South Kyushu of high temperature and humidity, and hoped for ecology life with the environment. The wind of land is felt, rain water is saved, and gardening is enjoyed. It is native life. The site is a residential quarter that extends on Shirasu Plateau near from the Kagoshima City downtown. The architects then thought native house(sustainablehouse) with the soil(Shirasu) as the material that formed this native plateau by the Shirasu block.

Shirasu has a lot of characteristics in other geological features like

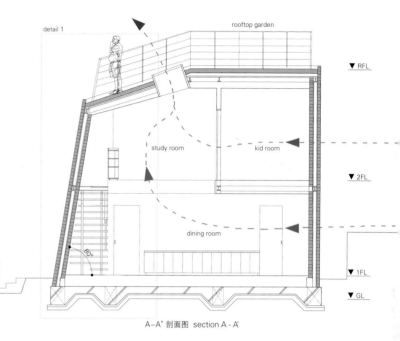

A-A' 剖面图 section A - A'

一层 first floor

二层 second floor

1 停车场	1. parking
2 入口	2. entrance
3 餐厅和厨房	3. dining and kitchen
4 起居室	4. living
5 卧室	5. bedroom
6 平台	6. deck
7 图书馆	7. library
8 书房	8. study room
9 卫生间	9. W.C.
10 儿童房	10. kid room
11 日光浴室	11. sun room
12 露台	12. terrace

项目名称：Shirasu House
地点：Kagoshima, Kagoshima Prefecture, Japan
建筑师：ARAY Architecture
首席建筑师：Asei Suzuki
结构设计：TMSD & Takashi Manda Structural Design
用地面积：228.9m²
总建筑面积：88m²
有效楼层面积：143.9m²
竣工时间：2013.7
摄影师：©Daici Ano (courtesy of the architect)

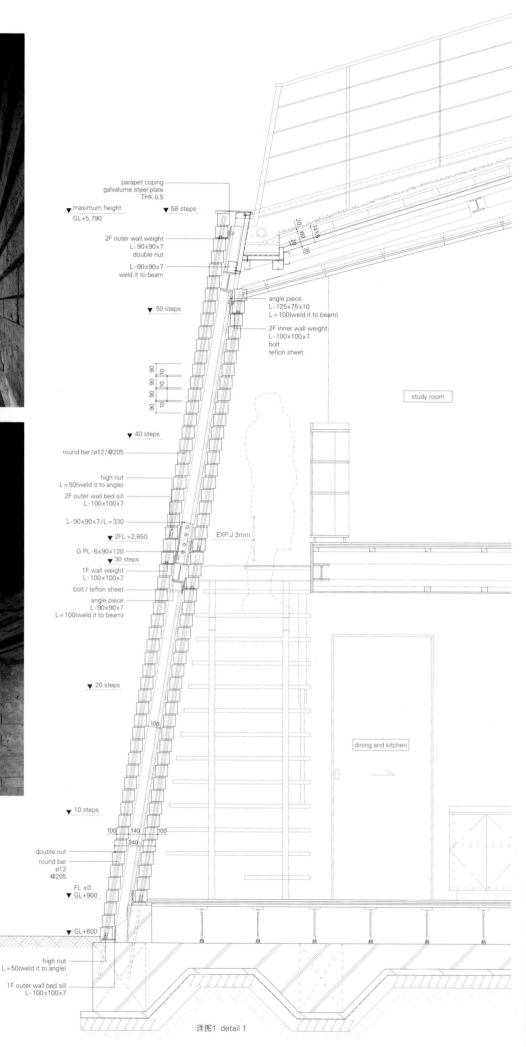

详图1 detail 1

fireproof, adiabaticity, the humidity conditioning, thermal storage, and lightness, etc.. While pressurizing and constructing it, the technology of a monotonous block in Shirasu for the pavement that had begun to spread in the city was made the best use of, and production with the block for the construction of Shirasu was tried for the first time. The outside wall block was inlaid with the raw ore of Shirasu to improve the adsorption and desorption of moisture to the inner wall block. At the same time, the character of this Shirasu appears as an expression of the memory. The house became a space wrapped in the soil like the cave.

This inside and outside mid-air layer double wall has reduced the thermal loading to the inside. In addition, the inner wall block surrounds like finish in any room of the house, and adjusts the indoor humidity. Therefore, the inside is chilly cool in winter, and warm in summer. It is a steady thermal environment throughout the year.

It proposed the energy performance of underground resources accumulated in the Shirasu Plateau. And House of Shirasu which reused the volcanic soil, is the new sustainable-house of environmental recycling.

In addition, the green roof and lightweight Shirasu block play the role of soil. Therefore, it is to reduce the burden of the roof load. The green roof of the Shirasu block, also can improve housing energy efficiency while positively impacting the quality of the urban environment.

And now, the resident is enjoying the life with familiar nature, stopping installation of an air conditioner and leaving control of a thermal environment to the Shirasu block's wall and roof.

ARAY Architecture

Okusawa住宅

Hiroyuki Ito Architects + O.F.D.A.

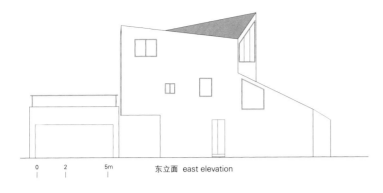
东立面 east elevation

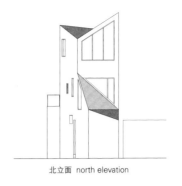
北立面 north elevation

南立面 south elevation

这座住宅是为一对夫妇及其孩子设计的，带有一间钢琴房，位于其父母家人居住的主建筑和车库之间。建筑师根据场地的轮廓，几乎确定了场地的平面，他们需要对主建筑内的起居环境，如光线、通风和私密性给予密切的关注。新住宅面向原有车库的屋顶，意在将其用作露台，并取代了与主建筑所面对的原有花园的连接，使每个家庭在某种程度上都能够独立生活。北侧，屋顶的高度受到了限制，使其没有对那里的窗户造成阻碍。南侧的体量则进行了抬高，以建造一个车棚，并且提供街道到原有花园之间的视野和通风。

斜对的弯曲屋顶有利于赋予室内合适的高度，以保持进入临近住宅的光线。天花板没有与地面保持平行，这种方式为钢琴房带来了非常好的隔音效果。

蝶形的屋顶由混凝土制成，将三楼分为两处松散连接的空间。建筑师发现通过建造分层的空间，能使房间看起来更加宽敞，而平坦的屋顶或者人字形屋顶似乎使建筑师们认识到房间的实际面积为27m^2。

混凝土墙体比木质墙体厚100mm，总体上来说，混凝土墙体应用在小型住宅时，能够增强紧密感。所以在设计之初，建筑师对于是否在这座狭窄的住宅内采用它持犹豫的态度，因为它能为钢琴房提供更好的隔音效果。但是考虑到小空间将作为舒适的壁龛来使用，建筑师还是坚定不移地使用了混凝土。家具也由混凝土制成，因为这里没有足够大的空间来放置现成的家具。因此，房间十分紧凑，但是非常温馨，如同一个窑洞一样。

在这些建筑构件中，如屋顶、包括家具的墙体，规格的相似性和材料的统一性使其间的分割变得模糊。建筑师意在建造一处空间，这处空间在被认知为某种功能构件和类似于桌子和天花板的象征之前，首先作为材料和形式被人们认知。

Okusawa House

The house for a couple and a child with piano room is planned on the lot between the main building, where the parents' household lives in, and a garage. According to the outline of the lot, the plot plan is almost decided. The architects needed to give close attention to keep the living environment of the main building such as sunlight, ventilation and privacy.

The new house faces the roof of the existing garage with the intent to use it as a terrace, instead of being connected to the existing garden directly where the main building faces, so that each household could live independently to some extent. The roof height is held down in the north part not to block up the window there. The volume is raised in the south part to provide a carport and also view and ventilation from the street to the existing garden.

项目名称：Okusawa House
地点：Setagaya-ku, Tokyo, Japan
建筑师：Hiroyuki Ito Architects + O.F.D.A.
结构工程师：LOW FAT Structure
用地面积：165.33m²
总建筑面积：38.40m²
有效楼层面积：95.32m²
竣工时间：2010
摄影师：©Jin Hosoya (courtesy of the architect)

1 起居室 2 厨房 3 餐厅
1. living 2. kitchen 3. dining
三层 third floor

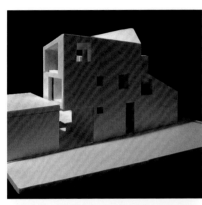

1 儿童房 2 浴室 3 露台 4 阳台
1. children's room 2. bathroom 3. terrace 4. balcony
二层 second floor

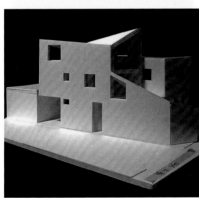

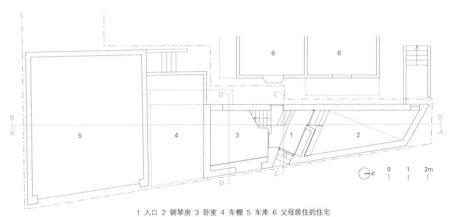

1 入口 2 钢琴房 3 卧室 4 车棚 5 车库 6 父母居住的住宅
1. entrance 2. piano room 3. bed room 4. carport 5. garage 6. parents' house
一层 first floor

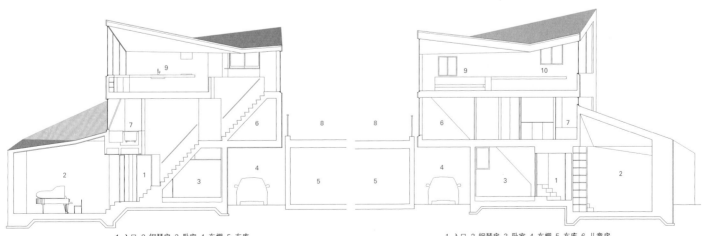

1 入口 2 钢琴房 3 卧室 4 车棚 5 车库
6 儿童房 7 浴室 8 露台 9 厨房
1. entrance 2. piano room 3. bedroom 4. carport 5. garage
6. children's room 7. bathroom 8. terrace 9. kitchen
A-A' 剖面图 section A-A'

1 入口 2 钢琴房 3 卧室 4 车棚 5 车库 6 儿童房
7 浴室 8 露台 9 起居室 10 餐厅
1. entrance 2. piano room 3. bedroom 4. carport 5. garage 6. children's room
7. bathroom 8. terrace 9. living 10. dining
B-B' 剖面图 section B-B'

1. polyvinyl-chloride sheet, rigid insulation foam t=50
2. exposed concrete
3. kitchen counter: water-repellent paint finish, concrete t=120
4. floor: mimoza flooring with a herringbone pattern t=18, structural plywood t=12
5. counter: tile=10, plywood t=12
6. floor: tile t=10, plywood t=12, double floor t=198
7. fibertree cement-board t=25
8. floor: waterproof membrane, trowel mortar t=40
9. waterproof membrane, fibertree cement-board t=30
10. counter: waterproof membrane, exposed concrete t=120
11. waterproof membrane, trowel mortar t=30
12. mosaic tile t=4
13. mirror

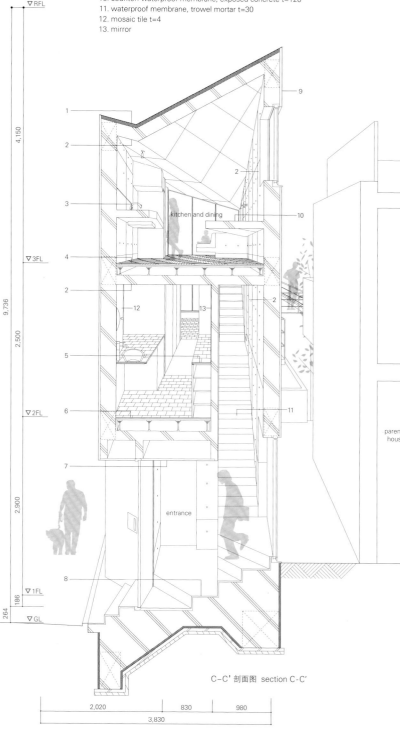

C-C' 剖面图 section C-C'

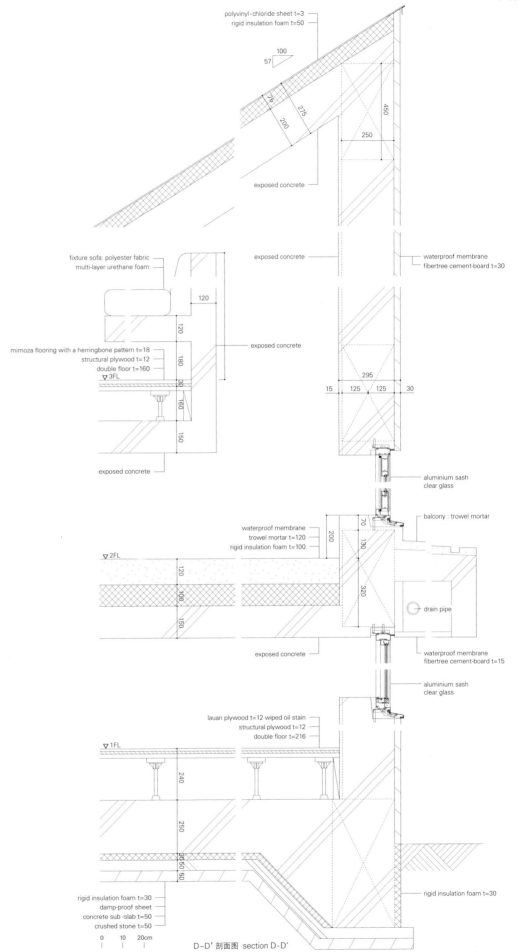

D-D' 剖面图 section D-D'

Bending roofs diagonally were convenient to give the building the height necessary inside, keeping the sunlight to the adjacent building. This method of making the ceiling not parallel to the floor, also brought acoustic effect to the piano room.

The butterfly roof made of concrete divides the third floor loosely into two spaces. The architects found that the roof would make the room more spacious making a layer of spaces, while flat or gabled roof seemed to make them recognize the actual size of this room of 27 square meters.

A concrete wall is 100mm more thick than a wooden wall, and in general it increases the sense of tightness when it is used in a small house. So at first, the architects hesitated to adopt it in this narrow house just because it had better sound insulation for piano. But they decided to use concrete affirmatively considering that small spaces would be snug niches. Furniture is also made of concrete, as there wasn't enough space to place ready-made furniture. As a result, the room became a compact but cozy space as a cave.

Both similarity of the size and unity of the material among the architectural components, such as roofs, walls including furniture, make the segmentation between them ambiguous. The architects intended to create a space which is recognized as materials and forms themselves, before they identify themselves as something with functions and symbols like a table and a ceiling.

Hiroyuki Ito Architects + O.F.D.A.

山外小屋

acaa / Kazuhiko Kishimoto

这是一个将一座住宅、一间办公室和一间小画廊的功能结合起来的结构。它由底层架空柱进行支撑，并且展现为一个漂浮的木头盒子的形象，这座建筑的特点就是它的外观。建筑与地面分离，且地面沿着前侧的街道呈下降趋势，因此建筑师利用了阶梯，使建筑没有与人流分开。因此，人们的目光落于此处，并且延伸至后面的树林中。阶梯通向一个开放的庭院。庭院是画廊的延伸，且作为一个室外画廊来使用。两座楼梯连接着居住区或者办公室。室外的另一个特点是它的天窗。街道的后移界切断的剖面完全覆盖了玻璃，以形成这个结构。天窗则产生了建筑内部光影之间的对比，并且对空间的未来操控具有重要的意义。

一层的前院呈阶梯状，并且与斜坡保持一致。在这里，人们可以直接坐下来。这个场地可以作为橡侧，即经典的日式住宅内可见的、狭窄的开放式木质走廊。庭院位于离小筑较远的后侧，且其旁边设有圆形长椅，它们均作为人们会客的场所。画廊内可以举办不同的展览，而前院和庭院则作为举办接待晚会的空间来使用。

Beyond the Hill

This is a combined construction of a residence, an office and a small gallery. Supported by pilotis and appearing like a floating wooden box, this building is characterized by its exterior. The building was set apart and freed from the land, which descends along the front side street, and a stepped approach was set up to prevent the building from blocking the flow of people. As a result, the gaze is drawn and expanded toward the trees stretching out in the back of the land. The stepped approach leads to an open courtyard. This courtyard is an extension of the gallery and also serves as an outdoor gallery. There are two stairways which connect to the residence or the office. Another feature of the exterior

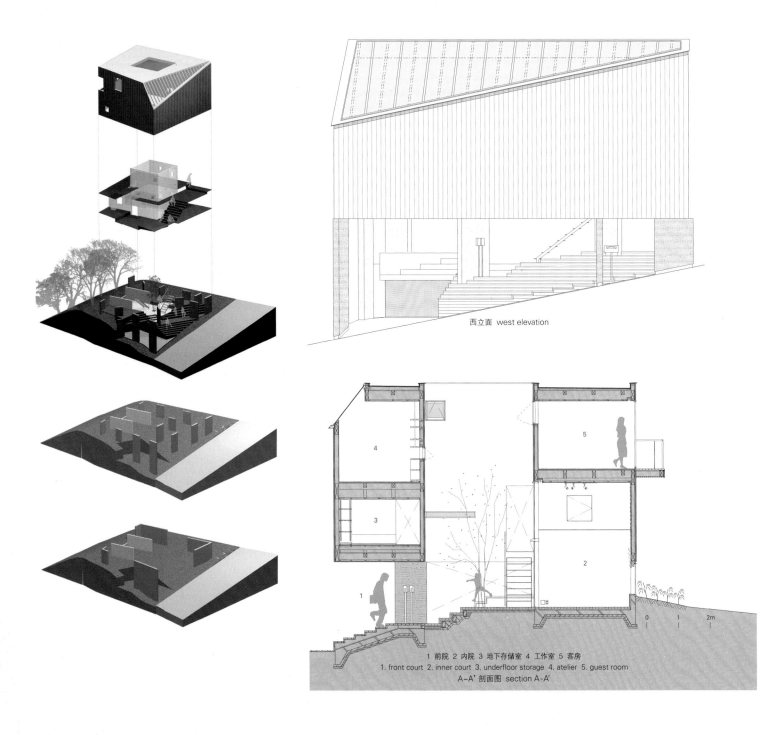

西立面 west elevation

1 前院 2 内院 3 地下存储室 4 工作室 5 客房
1. front court 2. inner court 3. underfloor storage 4. atelier 5. guest room
A-A' 剖面图 section A-A'

照片提供：©Ryogo Utatsu (courtesy of the architect)

is its skylight. The entire section cut off by the street's setback-line limit was covered with glass to create this structure. The skylight structure creates a striking contrast of light and shade inside the building, and has a significant meaning for the perspective manipulation of space.

The front court on the 1st floor is terraced in line with the slope, where people can directly sit down. It serves as an "Engawa", the narrow wooden open corridor seen in classical Japanese houses. The courtyard located further behind and the bench with a circular form next to it, were built as spaces where people can meet together. Various exhibitions are held in the gallery, and the front court and the courtyard are utilized as spaces for reception parties.

项目名称：Beyond the Hill
地点：Yokohama, Kanagawa, Japan
建筑师：acaa / Kazuhiko Kishimoto
项目团队：acaa
用途：residence, office, small gallery
用地面积：132.47m²
总建筑面积：79.39m²
有效楼层面积：158.39m²
设计时间：2011.10—2012.5
施工时间：2012.6—2013.1
摄影师：©Hiroshi Ueda (courtesy of the architect)(except as noted)

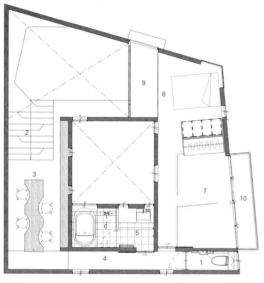

1 卫生间 2 休息室 3 工作室 4 通道 5 化妆间 6 浴室 7 卧室
8 客房 9 阁楼 10 阳台
1. toilet 2. lounge 3. atelier 4. passage 5. powder room 6. bathroom
7. bedroom 8. guest room 9. loft 10. balcony
二层 second floor

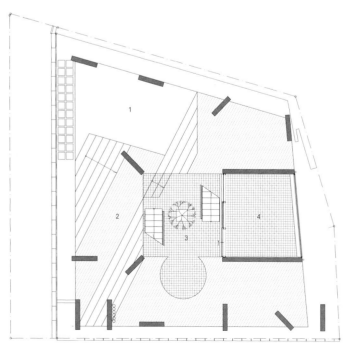

1 车库 2 前院 3 庭院 4 内院
1. garage 2. front court 3. courtyard 4. inner court
地下一层 first floor below ground

1 门廊 2 入口 3 套间 4 厨房和餐厅 5 地下存储室 6 空房 7 存储室
8 小厨房 9 卫生间 10 办公室
1. porch 2. entrance 3. inner room 4. kitchen and dining room 5. underfloor storage
6. spare room 7. storage 8. small kitchen 9. toilet 10. office
一层 first floor

日式城市住宅——传统与现代 Japanese Urban Dwell – Traditional Modernity

雪松住宅
Suga Atelier

建筑师稍稍提高了住宅所处的场地,使绿化区全部面北,并且能够俯瞰东侧的山脉以及西侧太阳的落山。尽管市郊区域人口密集,但是为了与这些自然特点相呼应,建筑师将北侧与南侧连接起来,通过木质的隧道来获取阳光。

建筑师采用非常普通的小规格木材(最大规格为120mm×180mm)作为材料,这些雪松大部分是附近生长的,因此对于建筑师建造的这座建筑来说,既简单又合理。所有的木框架在空间节点处由钢质螺栓(M12)进行固定,而不是使用机械连接和螺丝进行固定,因为它们既不稳固也不持久耐用,且无法隐藏。

此外,建筑师还采用干燥的木材,这些木材保持了其本身的天然颜色、气味以及强度。薄薄的保温墙覆盖着隧道,而在内部,胶合木材制成的墙体和地面分离开来,使每天的视野、光线和风流都保持通畅。南侧的阳光和屋顶天窗的光线照亮了木框架,使空间充斥着自由感,同时与树木整体和谐一致。

House of Cedar

The architects slightly elevated site of this house, making it face green towards north, slightly overlook the Mountains to the eastward and set sun to the westward. Despite densely populated in outskirt city area, responding to these natural good features, the architects connect northward and southward to get sun by wood-made tunnel.

Using very common small size wood (max 120x180mm) of cedar which is produced in bulk nearby, the architects make this house simple and rational. All wood frames are tightened three dimensionally by joints with simple steel bolts (M12), without using mechanical joints and screws which aren't persistent and durable, and keep all unconcealed.

Also the architects use wood of air-drying which keeps natural color and scent and strength. Thin insulated wall covers the tunnel and, inside plywood-made walls and floors are partitioned to harmonize daily convenience with the views, light and winds and so on. Sun from south-side and roof skylight, illuminate wood frames and reflect internally to fill over the whole space with light. The architects try to make a space which has a sense of freedom and companionship with trees. Suga Atelier

照片提供:©Yuko Tada (courtesy of the architect)

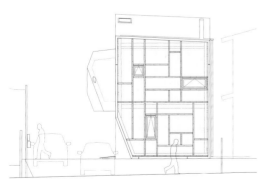

北立面 north elevation

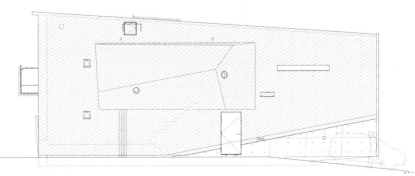

东立面 east elevation

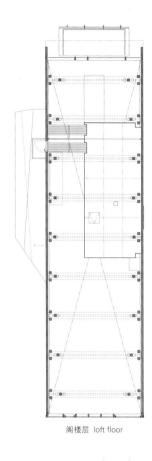
阁楼层 loft floor

二层 second floor

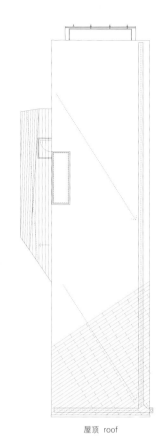
屋顶 roof

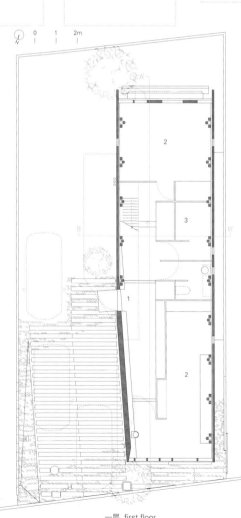
一层 first floor

1 入口	1. entrance
2 卧室	2. bedroom
3 存储室	3. storage
4 阳台	4. balcony
5 杂物房	5. utility room
6 厨房	6. kitchen
7 起居室/餐厅	7. living/dining

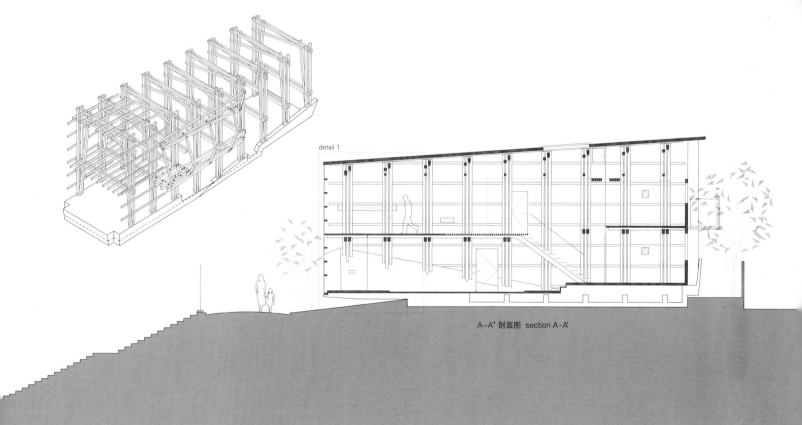

A-A' 剖面图 section A-A'

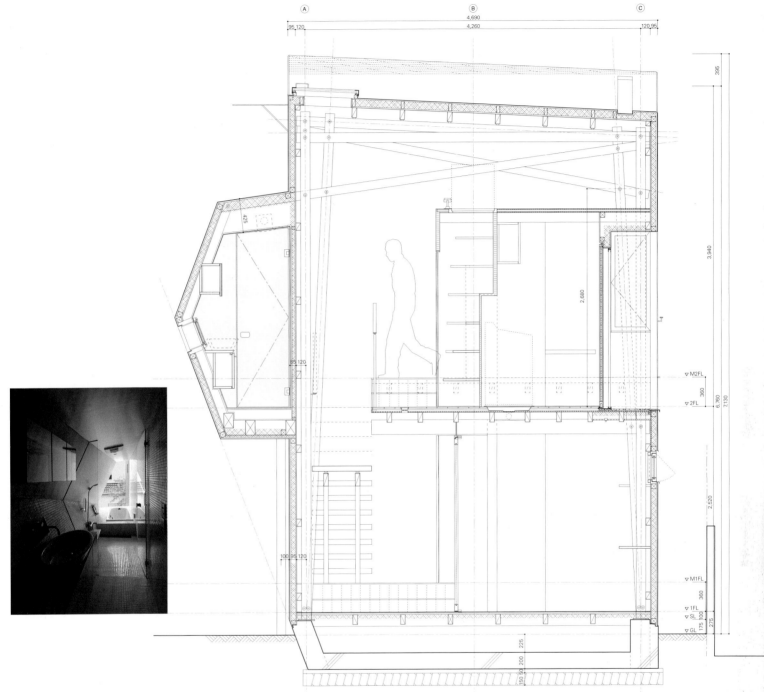

B-B'剖面图 section B-B'

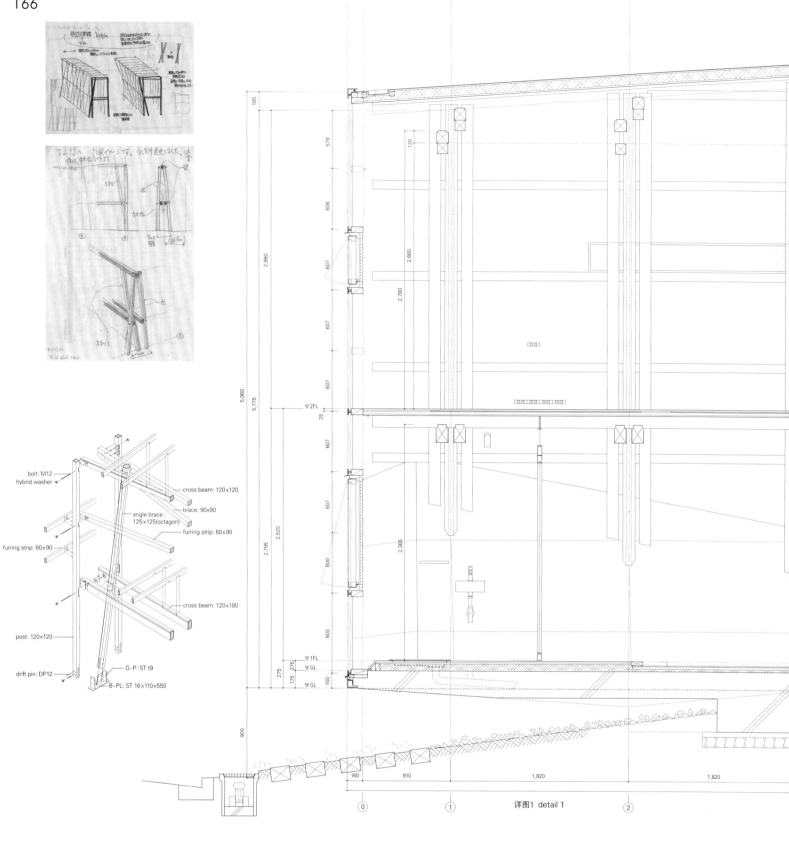

详图1 detail 1

项目名称：House of Cedar
地点：Osaka, Japan
建筑师：Shotaro Suga
项目团队：Shotaro Suga, Shintaro Amano, Tomoko Baba
结构工程师：Satoru Shimoyama
用地面积：205.1m²
总建筑面积：85.9m²
有效楼层面积：154.4m²
设计时间：2010
竣工时间：2011
摄影师：courtesy of the architect (except as noted)

螺旋之家

Hideaki Takayanagi Architects

设计的主要理念是"东京里的螺旋式门廊",以对打开(或封闭)城市环境的私密性进行挑战。日式门廊,即橼侧,在传统的日式住宅中是十分重要的,因此它也是所有空间类别中最带有怀旧气息的一种。近些年来,在东京,建筑场地的造价对于普通工人来说十分昂贵,因此,这一条件迫使建筑师放弃在住宅内建造门廊的想法。他们决定创造一个全新的橼侧,以节省空间,并且增加居住的优势。场地的地形十分狭窄,因此被迫形成一个立体的(螺旋状的)橼侧结构,该区域成为这座住宅的主要楼梯式房间。但是,在钢条外侧建造橼侧是一个绝妙的主意。钢条在室内和室外形成了许多隐蔽和隐藏的空间。这些空间内都可举办许多活动,如作为游乐室或者是用于休闲的日光浴室。橼侧的宽度较以往稍宽一些,所以居民能够将一些椅子或者家具放在此处,以装饰这一充满魅力的空间。

每层的楼板、螺旋结构和屋顶都是由4.5mm厚的铁板制成,它们都是由双层铁板制成的夹层构件,且规格适中,卡车能够将其从造船厂运出。日本的造船技术高度发达,墙壁和板材平且光滑,这是由焊接和热处理技术完成的。

这座住宅的质量非常轻,因此也足够柔软,能够抵抗地震。建筑没有使用厚厚的柱子。所有的部件都是由既柔软又坚硬的单壳体构件配置而成,如同铁船一样。

Life in Spiral

The main concept is "A Spiral Porch in Tokyo", challenging to open (or close) the privacy to urban environment. The Japanese porch, Engawa was essential to traditional Japanese house, therefore it is one of the most nostalgic space for all. Recently in Tokyo, the cost of building sites became too expensive for ordinary workers. Therefore, that made the architects to give up creating the porch

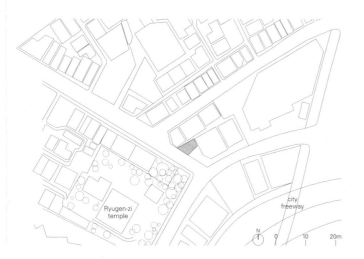

in the house. So they decided to create a brand-new Engawa that saves space and adds some benefits to live in. Quite a narrow site forced to form 3-dimensional(spiral shaped) Engawa structure forms the main stair room of this house. But this was a nice idea to create Engawa outside of the steel ribbon. The ribbon makes many shaded or hidden spaces inside and outside. These spaces afford many activities, such as the playroom or the sun room for relaxing. The width of the Engawa is a little wider than usual, so dwellers can put some chairs or furnitures to furnish this charming space.

Each floor slab, spiral, and roof are entirely made from iron plate of 4.5mm thick. These are made as some sandwiched components by dual iron plates. These are about the size so that trucks can carry them from the shipbuilding factory. Japanese shipbuilding technology is highly developed. Walls and slabs are very smooth and flat. Welding and heat treatment technique made them.

This house is very lightweight to be supple for earthquake. No thick columns are needed, and all parts were configured with supple and strong mono-coque body like iron ship.

Hideaki Takayanagi Architects

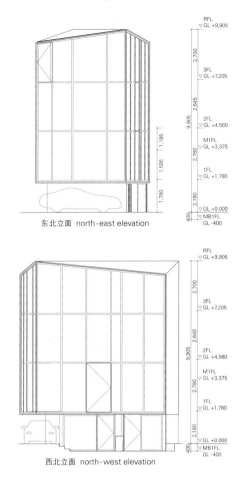

东北立面 north-east elevation

西北立面 north-west elevation

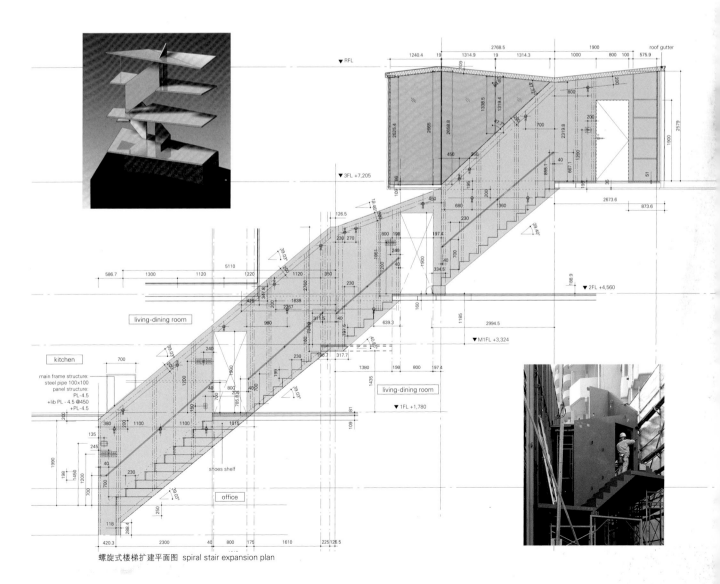

螺旋式楼梯扩建平面图 spiral stair expansion plan

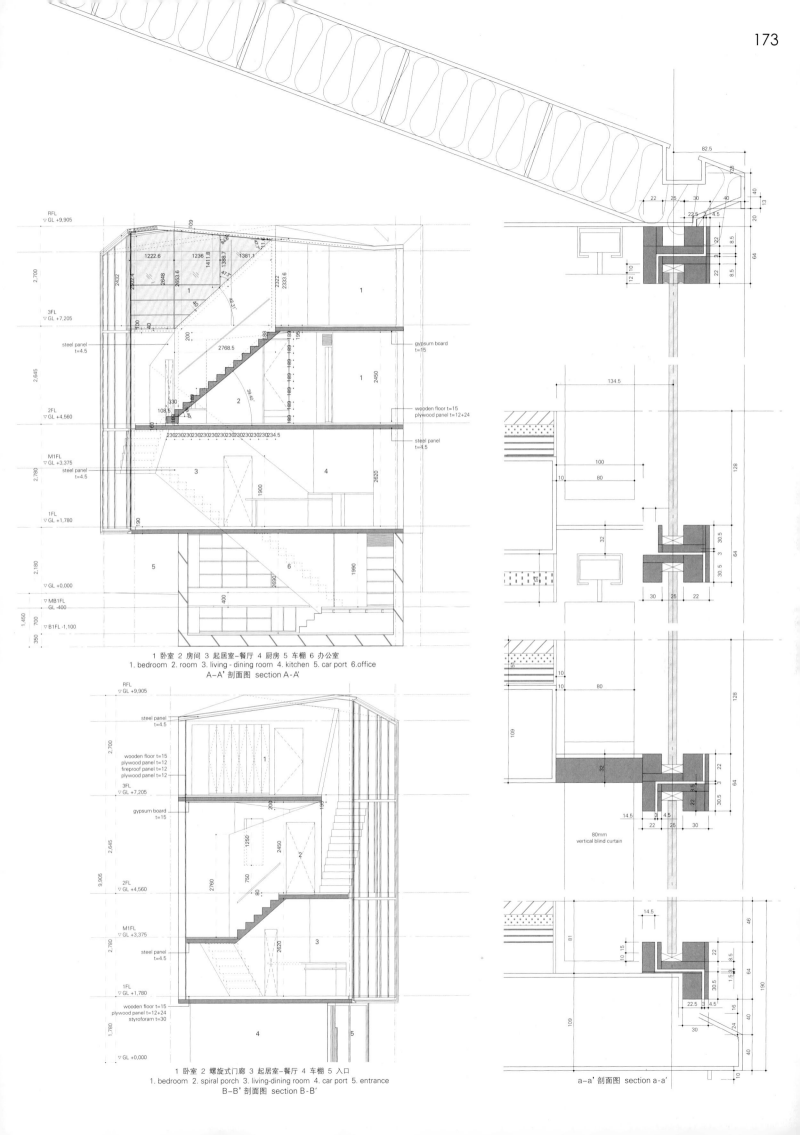

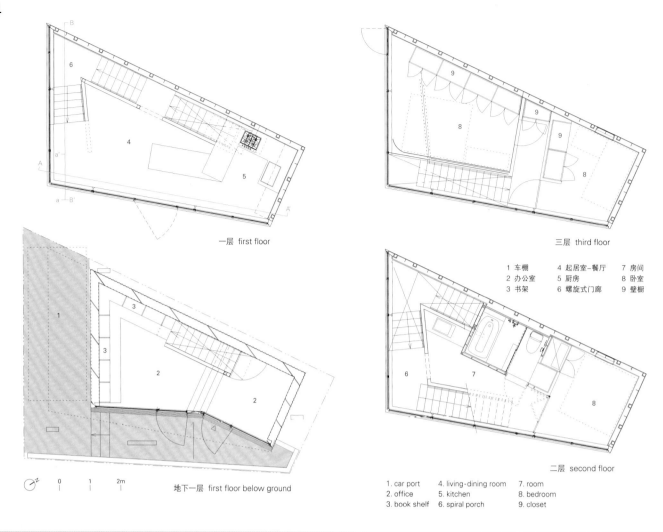

一层 first floor

三层 third floor

地下一层 first floor below ground

二层 second floor

1. car port   4. living-dining room   7. room
2. office   5. kitchen   8. bedroom
3. book shelf 6. spiral porch   9. closet

项目名称：Life in Spiral / 地点：Tokyo, Japan
建筑师：Hideaki Takayanagi Architects
用地面积：50m² / 总建筑面积：29.8m² / 有效楼层面积：109m² / 最大高度：9.9m
结构：steel panel made by 4.5mm thick plate
室外饰面：OP, St-PL T=4.5
竣工时间：2010.4
摄影师：©Takumi Ota

阳台之家

Ryo Matsui Architects

阳台之家位于一处住宅区，这里拥有一座低层住宅公寓综合楼以及一些带有小阳台的新建住宅，而建筑师设计了一座四层的钢筋混凝土建筑。

在东京，由于土地资源的缺乏，这座住宅只是占据了场地的边界区域。而在建筑所处的街区内，所有的建筑都拥有同样的场地条件。

建筑师研究了住宅区内阳台（小型且仅作为遮挡的花槽）的影响，其结果是阳台被设计为一个全新的街道景观。

为了完成这一目标，阳台的进深改为2m，成为道路的缓冲区，同时也作为屋檐发挥作用。建筑师也通过扩大阳台的地面来使这些花槽变得茂盛。

建筑师希望这些阳台不仅对住宅本身产生积极的影响，同时也能够创造一处全新的城市景观。

Balcony House

Balcony House is placed in a residential area which has a low-rise building apartment complex and new houses with small balconies, and is designed as a RC 4-store house.

In Tokyo, due to the lack of space, the house extremely approaches the site's boundary. The neighborhood this building is placed

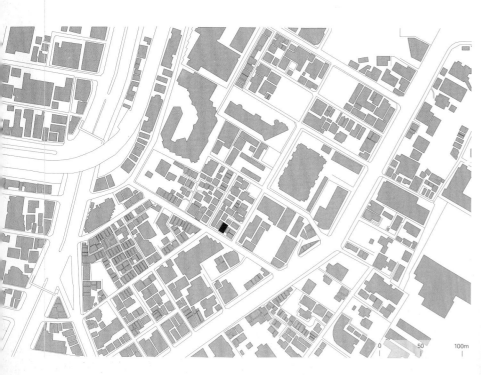

on has the same site conditions for all buildings.

As a study of the influence of the balcony in the residential area, in which balconies are very small and only used as planters for blindfolds, the result the architects want to approach is a new streetscape.

To archive this, the balcony's depth becomes 2m, which makes it the buffer area of the road and takes the function of eaves. the architects also allow plants to grow wild by increasing the floor of the balconies.

The architects suggest that balconies have a positive influence not only to the house itself, but also to a new cityscape's creation.

Ryo Matsui Architects

郊区住宅 suburban residence

城市住宅 urban residence

南立面 south elevation

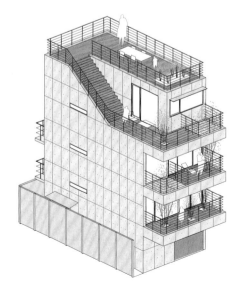

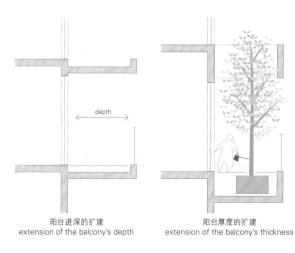

阳台进深的扩建
extension of the balcony's depth

阳台厚度的扩建
extension of the balcony's thickness

项目名称：Balcony House
地点：Minato-ku, Tokyo
建筑师：Ryo Matsui Architects
工程师：Akira Suzuki / ASA
建筑面积：118.5m² / 有效楼层面积：202.6m²
施工单位：Yoshinaga Kogyo Corporation
设计时间：2011.7—2013.2 / 竣工时间：2012.6
摄影师：©Daici Ano (courtesy of the architect)

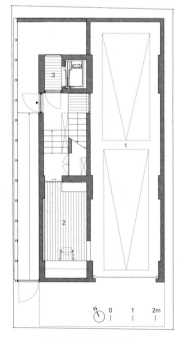

1 停车场 2 书房 3 电梯厅
1. parking 2. study 3. elevator hall
一层 first floor

1 儿童房 2 卧室 3 浴室 4 化妆间 5 阳台
1. child's room 2. bedroom
3. bathroom 4. powder room 5. balcony
二层 second floor

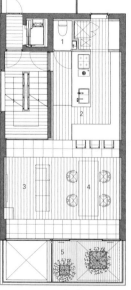

1 卫生间 2 厨房 3 起居室 4 餐厅 5 阳台
1. rest room 2. kitchen
3. living 4. dining 5. balcony
三层 third floor

1 化妆间 2 浴室 3 壁橱 4 卧室 5 阳台
1. powder room 2. bathroom
3. closet 4. bedroom 5. balcony
四层 fourth floor

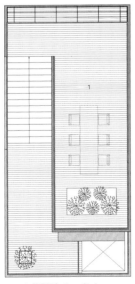

1 屋顶阳台 1. roof balcony
屋顶 roof

>>8
Spora Architects
Is a Budapest based architectural office, founded in 2002 by four partners: Tibor Dékány, Ádám Hatvani Orsolya Vadász and Sándor Finta[from the left]. They have studied at the Budapest University of Technology and Economics, Faculty of Architecture. Sourcing inspiration from almost everywhere, they see architecture as something fundamentally complex, where everything can be potentially interesting. They look at the profession from this point of view and recognize that social role of architecture is strongly increasing and the function is also changing widely and rapidly. They try to be active members of the Hungarian Contemporary Architecture Center(KÉK), an initiative of young architects and artists who are intended to create an independent cultural center open to all to strengthen community life and promote architectural education.

©Tamás Bujnovszky

Andrew Tang
Received his architecture degree at Institute of Design(IIT), Chicago in 1996. In the same year, he won the Jerrold Loebl Prize. Throughout his 20-year career, he has worked around the world contributing to many innovative and challenging projects in the field of Architecture & Urban Design. He has worked in the Netherlands on projects such as the new Central Station in Rotterdam serving as Architect, Urban Designer and Public Space Designer. Is currently a designer and founder of the design practice Tanglobe in Hawaii.

Michele Stramezzi
Received a master's degree in architecture and urbanism at Polytechnic of Milan School of Architecture in 2003. Has ever worked at a number of offices in the Netherlands, amongst others, de Architekten Cie, Erick van Egeraat Associated Architects and MVRDV as project leader. After moved to Beijing, China, he has been working as freelance architect and architectural consultant.

Heidi Saarinen
Was born in Finland, raised in Sweden. Has extensive teaching experience and project management, with specialism in experiential methodologies and spatial analysis in teaching and learning, encouraging students to interact with space through movement; questioning behavior and use of space. Is currently working on collaborative projects linking film, digital media, architecture and design, community and narrative space based in London. Is also a senior lecturer in Interior Architecture and Design at the University of Hertfordshire, and also eternal advisor at University of the Arts London and external examiner at Coventry University, UK.

>>86

ARX
Was founded in 1991 by Nuno Mateus and José Mateus. Nuno Mateus graduated from School of Architecture of Technical University of Lisbon(FAUTL) in 1984 and received a master of science in architecture and building design in 1987 from Columbia University in New York. Was a director of the department of architecture of Autonomous University of Lisbon between 2004 and 2007. Received a Ph.D. from School of Architecture from FAUTL in 2013. José Mateus also graduated from FAUTL in 1986. Is a guest professor at Superior Technical Institute(IST) in Lisbon and also president of the director's board of Lisbon Architecture Triennale.

>>72

MTM Arquitectos
Javier Sanjuan Calle and Javier Fresneda Puerto both graduated from the Polytechnic University of Madrid in 1992. They have been working as associates of MTM Arquitectos since 1997. Among the 17 built works, various congratulations through prizes and selections were given and they have obtained more than 37 prizes in competitions.
Javier Sanjuan Calle is a specialist in urban planning. He is teaching Architectural Design at the European University of Madrid. Javier Fresneda Puerto is a specialist in structure. He is also teaching Architectural Design at the Architecture School Alcala de Henares.

>>72

Xpiral Arquitectura
Javier Peña Galiano(b. Murcia, 1966), head of XPIRAL architecture studio, graduating from ETSAM in Madrid in 1992, worked for various firms, including Abalos & Herreros, before finally founding Xpiral in 1997. Is part of a new generation of architects who have refreshed the typical postmodern Spanish scene of the early 1990s. Has been honored with numerous awards like Architecture Prize Europan 6 and participated in several competitions and exhibition including the 8th Architecture Biennale in Venice.

©David Frutos

>>22
Team CS
Founded in 2003, is a cooperation between Benthem Crouwel Architects, MVSA Meyer en Van Schooten Architecten and West 8. The three offices are jointly responsible for the design of the new train station Rotterdam Central. They have cooperated together as one architectural firm. This collaboration and the chemistry between the offices have led to a plan that is more than the sum of its parts. Resulting in this plan of international standing that can compete with stations in London, Berlin or Barcelona.

>>52
Johnson Pilton Walker
is a Sydney-based design studio with major built works in architecture, planning, urban design, landscape architecture, interior design and exhibitions in Australia and internationally. The directors and key staff have been together since 1985. The work of the practice has been highly awarded across these fields. The White Bay Cruise Terminal project was lead by Paul van Ratingen[left], Director and Brendan Murray[right], Project Architect.

>>44
Dominique Perrault Architecture
Dominique Perrault gained international recognition after winning the competition for the National French library in 1989 at his age of 36. This project marked the starting point of many other public and private commissions abroad. In 2014, he delivers the DC Tower in Vienna, the tallest tower in Austria, an icon of the new business district, as well as the Grand Theatre in Albi, France. Received many prestigious prizes and awards, including the "Grande Médaille d'or d'Architecture" from the Académie d'Architecture in 2010, the Mies van der Rohe prize, the French national Grand Prize for Architecture, the Equerre d'argent prize for the Hotel Industriel Berlier and the Seoul Metropolitan Architecture Award as well as the AFEX Award for the Ewha Womans University in Korea.

>>100
TSM Asociados
Has 22 years of experience in the areas of Design, Integrated Development Project, Construction and Real Estate Management. Their main purpose is the provision of professional services as consultants, designers and implementers in civil area. Has established itself as a strategic partner in the development, growth and projection infrastructure of its various corporate clients.
General Manager, Alvaro Grimaldo Tremolada is in charge of Architectural Design area. Roberto Borda Sboto is in charge of Finance and Construction Management area as Administrative Manager. Alessandra Peña Roman is Senior Architect of Design area. All of them have studied Architecture at the University Ricardo Palma, Perú.

>>34
Ábalos + Sentkiewicz Arquitectos
is an international architecture office settled in Madrid, Spain and Cambridge, US, directed by Iñaki Ábalos[right] and Renata Sentkiewicz[left]. They have worked together first in Ábalos & Herreros since 1999, and from 2006 in Ábalos+Sentkiewicz arquitectos.
Iñaki Ábalos received a M.Arch and a Ph.D in Architecture from the School of Architecture of Madrid(ETSAM). Renata Sentkiewicz received M.S. from the Cracow Polytechnic School of Architecture. Have taught in Harvard GSD, AA, Columbia, Cornell, Princeton, ETSAM, combining the academic, professional and research activity. Received 18 first prizes in architecture competitions and currently developing projects in Europe, Latin America and China.

>>122
FORM / Kouichi Kimura Architects
Kouichi Kimura was born in Kusatu City, Shiga prefecture, Japan in 1960. Graduated from the Kyoto Art College in 1982 and established FORM / Kouichi Kimura Architects in 1994.

>>150
acaa / Kazuhiko Kishimoto
Kazuhiko Kishimoto was born in Tottori prefecture, Japan in 1968. Graduated from Tokai University in 1991. Established Atelier Cinqu in 1998 and changed the name of the firm in 2006. Currently, he is lecturing at Tokai University and Tokyo Designer Gakuin College. Believes that design activities are done by thinking not compliant to the trend but by conscience that is taken from the global perspective. Also thinks that detail and design lead from endemism is important than any other things.

>>142
Hiroyuki Ito Architects + O.F.D.A
Hiroyuki Ito was born in Saitama prefecture, Japan in 1970. Received a B.Arch in 1993 and a M.Arch in 1995 from the Tokyo University, Japan. In 1995, he has worked at Nikken Sekkei Ltd. Co-established O.F.D.A. Associates in 1998 and founded Hiroyuki Ito Architects in 1999. Has been lecturing at some universities including Tokyo University of Science, Tokyo Denki University, Ochanomizu University, Keio University and Nihon University since 2003.

>>168
Hideaki Takayanagi Architects
Hideaki Takayanagi is majored in architecture. Received B.S(1996) and M.S.(1998) from the Waseda Uni-versity, Japan. In 2003, he also received D.Eng. of Architecture Planning from his old school. Has taught at the Chiba University and currently lecturing at the University of Shiga prefecture.

>>114
Hiroyuki Shinozaki Architects
Hiroyuki Shinozaki was born in Tochigi Prefecture, Japan in 1978. Graduated from Kyoto Institute of Technology in 2000. After he completed a master's course in 2002 at the Tokyo National University of Fine Arts and Music, he has worked for Toyo Ito & Associates, Architects. Established Hiroyuki Shinozaki Architects in 2009.

>>160
Suga Atelier
Shotaro Suga was born in Osaka, Japan in 1956. Graduated with B.Arch from Kyoto Institute of Technology in 1980. Established Suga Atelier in Osaka city in 2001. Currently lecturing at the Osaka City University, Kobe Design University, Kinki University and Kyoto Prefectural University.

>>176
Ryo Matsui Architects
Ryo Matsui was born in Shiga Prefecture, Japan in 1977. In 2004, he graduated from the Tokyo University of the Arts with a Master of Fine Arts. Established Ryo Matsui Architects in 2008 in Tokyo, Japan and received Jury's award at the JCD Design Award in 2012. His design consistency of the spaces realized have been appraised and also awarded from number of organizations both internationally and domestically.

>>132
ARAY Architecture
Asei Suzuki was born in Shizuoka Prefecture, Japan in 1977. Graduated from the Tokyo University, Science Department of Life Science and Engineering Research in 2002. Has worked in Office of Kumiko Inui and the Hiroshi Nakamura & NAP Co.,Ltd. after the graduation. In 2009, he established ARAY Architecture

C3, Issue 2014.11

All Rights Reserved. Authorized translation from the Korean-English language edition published by C3 Publishing Co., Seoul.

© 2015大连理工大学出版社
著作权合同登记06-2014年第189号

版权所有·侵权必究

图书在版编目(CIP)数据

传统与现代：汉英对照 / 韩国C3出版公社编；王晓华等译. —大连：大连理工大学出版社，2015.1
（C3建筑立场系列丛书）
书名原文：C3 Traditional Modernity
ISBN 978-7-5611-9723-3

Ⅰ. ①传… Ⅱ. ①韩… ②王… Ⅲ. ①建筑设计－汉、英 Ⅳ. ①TU2

中国版本图书馆CIP数据核字(2015)第004812号

出版发行：大连理工大学出版社
　　　　　（地址：大连市软件园路80号　邮编：116023）
印　　刷：上海锦良印刷厂
幅面尺寸：225mm×300mm
印　　张：11.75
出版时间：2015年1月第1版
印刷时间：2015年1月第1次印刷
出版人：金英伟
统　　筹：房　磊
责任编辑：张昕焱
封面设计：王志峰
责任校对：高　文

书　　号：978-7-5611-9723-3
定　　价：228.00元

发　　行：0411-84708842
传　　真：0411-84701466
E-mail：12282980@qq.com
URL：http://www.dutp.cn